Remote Work Technology

Remote Work Technology

Keeping Your Small Business Thriving From Anywhere

Henry Kurkowski

WILEY

This book is dedicated to my sister Lilibeth, my brother-in-law Brad, and my far better half, Kirby. As I wrote this book, all three began battling cancer. The strength that all three have shown has been beyond measure. To say that they have been inspiring would be an understatement.

This book is also dedicated to all those who battle cancer and their families who fight alongside their loved ones in their own way. Keep fighting the good fight.

About the Author

Writer, entrepreneur, and author Henry Kurkowski is a pioneer in managed Wi-Fi technologies and digital customer engagement. He's an experienced founding partner in upstart technology companies with a long history in the telecommunications and SaaS industries. He has helped thousands of small businesses across the US leverage remote technologies and digital communications to automate operations, increase engagement, and boost the bottom line.

His insights on leveraging business partnerships and strategic brand marketing have been featured on *Forbes.com*. As an arts advocate and community-minded person, Henry has worked closely with arts and cultural nonprofit organizations as a volunteer serving on boards and subcommittees as well as professional nonprofit associations. He feels that there is no higher calling than to serve and give back to the community.

About the Author

Acknowledgments

I am eternally grateful for the opportunity to write this book but especially for the timing of it. This came at a period when I needed it most. As Kirby was diagnosed with acute myeloid leukemia during the shutdowns of the pandemic, the hospital visiting hours were limited. I would have spent every moment of the day at St. Vincent's Hospital if I were allowed, but COVID-19 precautions only permitted me to be there during more restricted visiting times. Working on this book gave my mind something to focus on when I was home alone, rather than thinking of terrible scenarios of all the things that could go wrong over the following months. Writing this book granted me a calm sea to sail where I could escape from the storms looming on the horizon, even if it were only for a few hours a day.

No one writes a book alone.

Writing this book has been made possible by the encouragement and assistance of many others. I owe my thanks to generous people for helping me along the way, a number of whom have also had a profound impact on me. Bradley Irvin has been a stabilizing force in many aspects of my life. I am thankful for the figures and graphs that he designed for this book. His formatting assistance and his organizational skills have been vital to me keeping my deadlines. But above of all of that, Brad's presence in my world allows me to remain compos mentis. Special thanks to Mason Dutch Yochum, who said that he would be there for me to help in any way in my home or in my life should I take on this project while managing the worries of Kirby's leukemia, my business responsibilities, and everyday life. I am thankful for Mason's help with transcriptions of the interviews in this book, and I am thankful as well for him being there to talk to on the days when news of setbacks would arrive from the team of doctors at the hospital.

Kristian Andersen has been my friend throughout a number of interesting stages of my life, and he is a busy man. I am grateful for him taking the time to write the foreword for this book and for being a confidant. Big gratitude to Brandi Andersen, who hopped into her van and was the very first person on the scene when word of Kirby's condition came around. She came to hug me and let me know that I was not alone. Knowing that you are not alone when isolated from others is a big part of this book. That loving act of Brandi's was a source of strength and hope. A huge thank you is given to my friends and compatriots in arts advocacy, Laura Glover and Mark Edgar Stephens, for their words of wisdom and expertise contained within this book. I am also thankful for them both being fountains of good information, generosity, and inspiration.

A great deal of gratitude goes out to another fierce arts advocate who also happens to be a talented actor and marketer, Paul Hansen. Thank you, Paul, for being kind, thoughtful, and well-networked and for introducing me to the great folks at Wiley publishing, in particular Senior Content Acquisitions Editor Kenyon Brown, who has been a guiding light through this process and has gone to bat for me several times. Thank you to my editor and project manager, Robyn Alvarez, who has the patience of a saint and has been able to work through my formatting mishaps while offering excellent input and suggestions along the way. I am grateful for the work of Christine O'Conner, the managing editor for this book. Thanks as well to Saravanan Dakshinamurthy for his work on the book and its content design. I am also grateful for the input, knowledge, and experience of Senior Managing Editor Pete Gaughan.

Special thanks to Laetitia Smith and Kristin Treat for working with me and helping put me in touch with the engineers from Nintex. Thank you to author Benjamin Spall for his correspondence and assistance.

I have nothing but sincere appreciation and gratitude for the following entrepreneurs who took the time to share with me the stories and reflections on their experiences of taking their companies remote. Thank you to Matt Smith, with whom I went to high school on Long Island in the days of wearing double Polo shirts with the collars popped up. Although we didn't have the chance to be friends back then, we have made real connections with each other remotely, and I am glad to now be able to call you my friend. To my fellow Indianapolis entrepreneurs and business leaders, Leslie Murphy, Jeb Banner, and Trevor Yager, the three of you have all touched a great number of lives in Indiana and have each helped make our city a better place in which to live, work, and play. Wendy O'Donovan Phillips, I offer extra thanks for your follow-ups and your introductions. David LaRosa, JP Holecka, Brian Handrigan, Scott Baradell, Marc Aptakin, and Marla DiCarlo, I thank you each deeply and feel honored that you chose to open up to me and were able to express the details of what you each went through during trying times. You are fine examples of the right way to lead a company and how to treat people. Thanks to Max Yoder, Jennifer Thoemke, and Christopher Hooper, who also took the time to speak with me and share their experiences.

Unending gratitude goes to Kirby, who each day helps me to become a better person than I was the day before.

Contents at a Glance

Contents

Foreword

When Henry reached out to me about the prospect of penning an introduction to this book, I enthusiastically agreed. After all, as a founding partner of High Alpha, I had recently gone through the process of overseeing our transition from a traditional office environment to a hybrid/remote-first organization. Early on, during the first phase of the COVID-19 pandemic, my partners and I made the decision to fully transition our team to a remote-first work environment. Throughout the process, I took a very hands-on approach to shaping our strategy, specifying our tech and software infrastructure, establishing our safety protocols, hardening our security procedures, maintaining our winning culture, and effectively managing communication with our team. I know firsthand what a monumental task it is to quickly transition an entire organization from on-premises to remote work while maintaining productivity and minimizing disruption. We ended up managing the transition remarkably well, but had *Remote Work Technology: Keeping Your Small Business Thriving from Anywhere* existed when we began our process, it would have saved us an enormous amount of time, energy, and second-guessing.

It is of course an understatement to say that we are living in unprecedented times. Mega-trends such as the rapid adoption of cloud technology, the meteoric growth of e-commerce, the work-from-anywhere movement (fostered by access to nearly ubiquitous broadband and 5G), and the advent of the gig economy are reshaping how we do business. When those trends coalesce with a pandemic, political unrest, and society's demands for greater equality and more equitable access to opportunity, it creates both opportunity and perils aplenty. Strong leaders remain focused on opportunities for business transformation and growth, but the road to capitalize on those opportunities is often littered with pitfalls, and the potential for missteps is very real. We are on the precipice of a massive shift in how work gets done, and remote work will play a central role in that shift.

In our current context, there may be no more salient topic for a business leader than understanding how to rapidly transition to a remote work environment. After decades of flirtation with remote/hybrid work environments, every business in the world now has to confront the very real possibility that much of their workforce will transition to remote work. And up until now, there really was no road map for businesses to rely on as they rapidly remade themselves into remote-first organizations. Thanks to Henry and *Remote Work Technology*, that map now exists and is accessible to anyone who might require it.

When making the shift to a remote work organization, there is much to consider. Frankly, the list of decisions is downright overwhelming. Henry's approach, detailed here, demystifies the process while providing a flexible structure that can be applied by any business—large or small, traditional or progressive.

As is always the case, business leaders attempting to navigate uncharted waters would do well to seek out and implement the advice of those who have seen it before. I can think of no finer guide to help the reader navigate the quickly evolving landscape of remote work than Henry Kurkowski. Henry has spent the bulk of his career working at the intersection of technology, culture, and organizational transformation. His sincere passion for unlocking the hidden value in small businesses and his unparalleled ability to translate the esoteric into practical language is on full display in *Remote Work Technology*. Henry has written the definitive guide to "going remote" and, if you read its contents, I'm confident it will position your organization to thrive in the years ahead.

Kristian Andersen
Partner, High Alpha
www.kristian.vc

Introduction

My first time working remotely was in 2002. I was working in commercial finance, and I had a client that was busy buying up hotels in South Florida to then convert them into condominiums. The client had a constant flow of properties that he was readying to purchase, and time was an important factor to his business plans. At that time, I was traveling back and forth from Fort Lauderdale to Indianapolis once a month. DSL was just rolling out to smaller cities, and I was lucky enough to have access to faster Internet than the dial-up speeds that many homes and small businesses still used. Smartphones were not yet a thing, but personal digital assistants (PDAs) such as the PalmPilot were all the rage with businesspeople on the go. My main tools were email, an eFax account, and a Sony Ericsson mobile phone, which I still have to this day. With these high-tech tools at my command, I felt unstoppable. I came to fully appreciate that I didn't need to be in the office to work with equity investors, lending institutions, or commercial real estate agents.

I became hooked on the freedom that the virtual office provided, and I saw the advantages that the virtual office can provide an organization. I believed in this so much that I've designed the companies that I helped cofound to operate with remote workers and a decentralized office in mind. Our teams are distributed across the United States, and many of our developers are working from other countries. We extend that remote capability at the client level as well. We use our cloud dashboards and the automation features of our software to remotely manage thousands of devices at our clients' locations on a daily basis regardless of where they are located. However, most small businesses are not designed from the get-go to operate with distributed teams or collaborate remotely. That fact became obvious in 2020.

The outbreak of the coronavirus caused us to avoid other humans and take shelter in our homes. We as people were not fully prepared for it, and neither were an overwhelming number of small businesses. There was a great deal of struggling in the beginning weeks of the shutdowns. Mistakes were plentiful, and stress was at an all-time high. But it doesn't take a global crisis to force a business to suddenly go remote. Hurricanes, fires, and other disasters can catch companies unaware. Many times, the company owners and managers are not sure what to do to keep the company going if they suddenly can't work from their HQ. This book is written to help with that problem and offer real solutions. Just as my companies empower our clients with SaaS technologies and the automation that comes with them, my desire is to help set up more small businesses for their own success through this book.

As a tech guy, I am also a sci-fi buff. So, I am thrilled to quote one of the great leaders of the sci-fi world, Captain Jean Luc Picard: "What we do in a crisis often weighs upon us less heavily than what we wish we had done, what could have been." I want you to be in a position where you don't need to worry about what you wish you had done. I don't want you to have to be burdened with what could have been. I want you to be able to thrive and have your people feel secure when a crisis strikes your company. That is the purpose behind this book.

The information and best practices within are not limited to a crisis situation. This book is a guide for startup companies, for independent contractors, and for business leaders who want a road map to make their company operations virtual, or even to create a hybrid of traditional and distributed teams. Being able to run your company remotely is about the power of choice. It is the freedom to take your company further and be able to have your teams do meaningful work from more places without being tethered to any physical space.

With the right technologies, anyone can work remotely. But it takes more than tech to be successful at working remotely, and certainly it takes more than tech to successfully lead a remote team. That's why I discuss in detail the importance of healthy company cultures that empower their people through trust and transparency. There are some downsides to working remotely. Many workers during the pandemic spoke of Zoom fatigue, feelings of isolation, feelings of being disconnected from the company, and lowered employee satisfaction. It is wildly important to keep these very real problems in mind when taking the team remote.

This means adjusting management styles, altering how productivity is measured, and enabling higher levels of engagement.

"There are no ordinary moments." That is one of my favorite quotes by Dan Millman. It certainly is an apt one for the year 2020. Nothing about 2020 was ordinary. But in that year, business leaders were able to garner deeper insights about virtual teams, learn how to drive the success of these teams, and thrive in the remote working world. By the time you reach the end of this book, you will be able to do the same.

Henry Kurkowski
www.theremoteworkbook.com

CHAPTER

1

You Can't Go to the Office: Where Do You Go from Here?

Indianapolis, April 3, 2006 – Public safety officials are still concerned about Sunday's storm damage to the Regions Bank tower downtown. The straight-line winds blew out windows on several floors and peeled away parts of the building's facade.

While the wind is responsible for the initial damage to the city's third tallest building, wind will keep the area around it closed for at least another day. Channel 13 Meteorologist Jude Redfield says the wind speeds are expected to be up to 25 miles per hour through Tuesday afternoon. That means falling glass and debris from the Regions building are still a big safety concern.

Three sides and 16 floors of the building are damaged and it will be a while before the tower is repaired. Spokesperson Myra Borshoff says that is bad news for tenants.

"For the foreseeable future, this building will not be occupied...Right now, our focus is securing the building and making it safe."

In the short term that means protecting pedestrians and drivers from falling glass and metal. And, Borshoff says, that means keeping the area around the tower off-limits.

"I would say for at least for the next few days, those streets will not be available and you should probably plan to leave 15 to 20 minutes early."

Mayor Bart Peterson agrees. "When we are sure nothing is going to fall off the buildings, then we will reopen the streets."

But the businesses and their employees who work at One Indiana Square will have to wait longer. Getting back to business for them means finding temporary office space elsewhere. Susan Matthews, with the building's management, says they are doing their best to help. "We are in the process of contacting tenants, assessing their needs, and helping them find space."

Building owner Mickey Maurer says they'll do what is necessary. "That's part of being local. We're here. We told those tenants we'd stand behind them, and we've got a plan, we've got an army put together, and we will stand behind them."

Hotels and office buildings in the downtown area have reached out, offering their extra space. One Indiana Square has about 30 tenants and 1,000 employees. Among those are accounting firms who are eager to get back to work with tax day just two weeks away.

Lynsay Clutter-Eyewitness News, Area around Regions Tower to stay closed, WTHR, April 4, 2006

Disaster Strikes

Imagine that you are awakened in the wee hours of the morning to be told that there was a fire in the middle of the night. Although nobody was harmed, your workplace has been severely damaged and is unusable. What is the first thing that pops into your head? Does your mind dart to worry about the various company files and assets that were in the office? Is the anxiety of upcoming project deadlines lumping in your throat? Do you race your thoughts around trying to figure out how to keep the team working and the company operational?

All of these are valid concerns, and there is no correct order of priority. This type of scenario has happened countless times around the world. Sudden man-made and natural disasters have interrupted the flow of business for centuries. In the preceding news report from WTHR in Indianapolis, you can see a good example in which roughly 1,000 workers of over 30 businesses were suddenly displaced without warning.

How Long Will the Crisis Last?

The building mentioned in the Eyewitness News story from April 2006 is the third largest building in Indianapolis. The storm that hit the downtown area was identified by meteorologists as a derecho. A derecho is a series of fast-moving wind storms or thunderstorms with powerful straight-line winds that can rival the force of hurricane winds. That storm hit downtown Indianapolis late on Sunday, April 2. By Monday, businesses were told that the building was unstable and unsafe for anyone to gain entry. In fact, the surrounding streets were also closed off due to falling debris. Over the course of a weekend, dozens of businesses were turned upside down and displaced.

Nearly three months would go by for many of those tenants who were without a place to work and were unable to access their company assets or client files. That Monday morning, the business owners and managers were coping as best they could. Many were hoping that they would be out of their office for just a few weeks at most. Few, if any, realized that it would be months before they could return.

WTHR had a follow-up report when many workers returned to the building to open up their offices and called it a kind of homecoming. That second report was published 10 weeks later on June 10, 2006. They also reported on that day that there were still many offices that remained under repair and unable to open for business.

COVID-19

The 2020 pandemic forced businesses around the globe to close their offices. Most of those businesses either figured out how to work remotely or stopped operations for a period of time. People who thought that such a long-term interruption of day-to-day business couldn't happen to them suddenly found themselves dealing with it as their new reality. This begs the question, how many businesses were ready for an emergency in which the whole company needed to suddenly work remotely?

BEST PRACTICES

Get a company continuity plan in place for your business to make sure your company can survive and thrive in case of various types of crises.

Prepping the Whole Team

I have worked on the board of trustees for a nonprofit organization where I and several other board members were tasked to come up with a detailed continuity plan for the organization. A continuity plan is a vital document that outlines various courses of action to help a company not only survive but continue to thrive when disaster strikes. Of course, such plans are something that you want in place *before* your interruption in business occurs so that you are not left scrambling or arguing over the correct decisions to make in the midst of your crisis event.

With nonprofit or for-profit businesses, putting such plans in place demonstrates good stewardship and that you take your responsibilities to the stake holders of the organization seriously. Part of that plan needs to be a remote work contingency.

A large portion of the workforce already has the ability to work from home due to the proliferation of broadband. In fact, according the US Census Bureau, just prior to the COVID-19 crisis, about one third of workers had worked remotely and roughly half of information workers are able to work from home. However, being able to do work from home on a long-term basis and actually doing so are very different things. Historically, there has been some pushback from management on over-seeing a remote and distributed workforce.

Management Work Policy

According to the International Workplace Group, an international provider of office space, business lounges, and conference centers, more than 50 percent of companies surveyed in 2019 did not have a remote work policy in place.[1] When asked the reason, they simply cited that it was a long-standing company policy. In other words, they prefer managing their people in person.

Many times, the reason companies avoid working remotely is due to the company culture. There is a camaraderie that comes with being in the same room and bouncing ideas off each other. There is an energy in that room that cannot be duplicated in a video conference.

Other times, remote working is frowned upon due to employer preferences in management styles, and sometimes it is simply that management does not like change. In working with thousands of small businesses across the US, I have seen clients at all management levels resist introducing new technology and new methodologies. Many times, that resistance to change comes in the form of "that's just how we've always done things."

From the experience of managing my own companies, I certainly understand that initial resistance. Changing processes is a disruption, and it can be uncomfortable to learn new methodologies in day-to-day operations, especially if it makes a dramatic alteration in how a company manages its business.

Some leaders believe the best way they can manage people and performance is in person. There is trepidation in trying a new office format, and there are concerns about a potential loss of productivity during the learning curve of running a company in a new way. But when disaster strikes, we don't get the luxury of having many choices. When it comes time to adapt workplace operations in an emergency, the situation quickly becomes sink or swim.

IBM and Yahoo!

In the late '90s and into the early 2000s, IBM was making headlines for its progressiveness in allowing a large portion of its workforce to go remote. It added that the profitability of the company had greatly increased due to this strategic move. Having workers go remote reduced overhead by getting rid of 78 million square feet of office space. The company sold off 58 million square feet of that space for $1.9 billion. By 2009, IBM had 40 percent of its employees working remotely and was considered a pioneer in a remote working revolution.

However, in May 2017, the *Wall Street Journal* reported that IBM was giving its remote work employees a choice: either move back into one of the company's regional offices or leave the company.[2] The move was intended to increase the speed of productivity and create better collaboration between teams.

In 2013, the new CEO of tech giant Yahoo! made headlines when a company memo was leaked. In it, Marissa Mayer abruptly ended the renowned company perk of working from home. Again, collaboration was cited as the main reason for this sudden switch. Here is a part of the memo:

> *To become the absolute best place to work, communication and collabo-*
> *ration will be important, so we need to be working side-by-side. That is*
> *why it is critical that we are all present in our offices. Some of the best deci-*
> *sions and insights come from hallway and cafeteria discussions, meeting*
> *new people, and impromptu team meetings. Speed and quality are often*
> *sacrificed when we work from home. We need to be one Yahoo!, and that*
> *starts with physically being together.*[3]

When these events happened, many large companies followed suit and ended or discouraged work-from-home policies. Companies such as Honeywell, Best Buy, Aetna, and Bank of America all changed their telecommuting policies in order to have better control over the workday and have more face-to-face collaboration. Does all this mean that remote working is detrimental to a company's efficiency? The short answer is that it can be, but it does not have to be that way. Certainly not anymore.

In the second half of 2020, HR consulting and benefits firm Mercer surveyed 800 employers to ask about the productivity of workers who went remote due to the health crisis. Ninety-four percent of the employers surveyed said that productivity was the same as it was prior to the pandemic, with 27 percent of those reporting that productivity was actually higher. On top of that, 73 percent reported that one quarter or more of their employees would likely remain as remote workers after the pandemic.

In November 2020, a study commissioned by tech giant Microsoft showed similar findings. Boston Consulting Group and KRC Research conducted the study and polled 9,000 employees and managers from large companies across 15 European markets, asking about productivity of remote workers during the pandemic. Thirty-nine percent reported that the newly remote workers were as productive as they were before going remote, 34 percent reported that they were somewhat more productive, and 10 percent stated that they were significantly more productive. So what's the difference between 2017 and 2020 when it comes to the idea of remote work productivity?

Consider the leaps we have taken in residential broadband speeds since 2017. Today there are gigabyte home internet routers and network switches. High-speed cellular networks and 5G can act as backup connections to help keep people working. The technology for remote work today far exceeds what was available in 2017. Another point to take into consideration is that there has been a wake-up moment in management styles due to the COVID-19 pandemic. There is now a deeper understanding of how to create stronger collaborations between teams and how to better use the technology that is available.

Top Concerns of Management

Some of the first concerns managers and team leaders will have will be about functionality, such as how to get the team to function cohesively while separated and how to manage their newly distributed work force. Both of these concerns can be overcome if you don't fight them. Let's take a look at managerial obstacles.

According to 2019 a survey by OWL Labs,[4] some of the concerns that managers have about a remote workforce include diminished productivity, loss of employee focus, and even concerns about the long-term career goals of employees (Figure 1.1).

Top Concerns of Managers of Remote Teams

82% Reduced employee productivity

82% Reduced employee focus

75% Reduced team cohesiveness

70% Maintaining company culture

67% Employees overworking

65% Employees' career implications

Figure 1.1: Top concerns of managers of remote teams

Source: Modified from Owl Labs, State of Remote Work 2019, September 2019.

The truth is that a good number of managers have some anxiety about managing remote workers. Managerial skill sets get honed by repetition. For years, those management skills relied on large amounts of small face-to-face check-ins and quick meetings with staff members. Many people in leadership positions still gauge dedication by who stays late and who comes in early.

That style of management makes the move to remote working difficult and is the reason many companies chose not have a remote working policy.

I've had the unfortunate experience of working under managers who were always on the lookout for slackers. Everyone knew that if the manager was coming around, you had to make yourself look busy. If you were not doing something productive when they came by, they would find something for you to do or you would get sent home. That was their management style, and cracking the whip was something that permeated into the company culture.

That kind of management style is the antithesis of a productive remote work environment. Vocalized expectations, setting productivity milestones, and creating a culture of trust will go far in the switch to a distributed workforce. We will go into detail on those points in later chapters.

Do Not Fight the Tide

If you go swimming in the ocean and find yourself pulled away from shore by the riptide, the best advice from experts is twofold: Do not panic and do not fight the tide.

Do not panic is sound advice for any situation, but *do not fight the tide* is the key to survival. The reason why is simply put: you can't win that fight. The undertow that is pulling you away from the safety of the shore is strong and consistent. If you try to fight the tide, you will feel like you are making some headway, but in the end, any forward motion you get will be incremental and won't get you to shore. The only thing your struggle will accomplish is a severe depletion of your resources.

To survive, you need to go with the flow. Let the riptide take you out and when it stops pulling at you, then you act. You first swim parallel to the shore and out of the current that brought you to where you are now. Then you are in the proper position to get back on track to your goal of getting to the beach. You need to take the same approach if you suddenly find yourself in circumstances where you don't have a space for your employees to work.

For this book, I interviewed a number of business executives whose companies suddenly shifted to working remotely. Some of them resisted the need to go remote in the beginning. I've asked those people what they might do differently should they again find themselves in similar circumstances. The responses from them were all similar. Although they feel they adapted quickly to their situations, they would most likely accept their need to move to remote work earlier on.

One of the business leaders whom I interviewed is David LaRosa. Below is a segment of our discussion where he describes the series of events that caused his company, Verso, to go fully remote. Accepting the circumstances instead of fighting them removed stress and the pressure of timeliness and ultimately made his agency and its company culture emerge stronger from it all.

> It was 2018, and at the time Verso was about 28 people. We had a lovely space in downtown Santa Monica overlooking Broadway, one block from the train and six blocks from the beach. It was a nice space.
>
> It was a very rainy week in February. The first call came on a Tuesday morning—our largest account would be ending the relationship within the month. We had been working together for four years, and at the time we had probably 8 to 10 people working on that account daily. It rained a lot that day. The second call came on Wednesday night. I was at the office late, and our second largest account needed to end its relationship within 30 days as well. It rained a lot that day, too...
>
> The first step, clearly, was layoffs. We had 28 people in January, and by March we were down to 11 people.
>
> The next step was getting rid of our space—it was overhead we didn't need. I found a space down the street that was much cheaper, probably because of the large construction project happening right behind the space. I think they are still working on it today, but it was very nice for the price. However, when I went to have the last walkthrough, the space was flooded, and the tenants next door—a well-known editing house here in LA—were carrying their computers out in the rain. They said, "Don't lease this space. It floods every time it rains, and the landlord is a crook."
>
> So, I paused because we had two weeks left on our old lease. During those two weeks, I suddenly had to fly to South Carolina to deal with a family emergency. As the lease ran out, my working remotely from South Carolina started a company trend. The whole team started working from home rather than take the train or drive two hours to the office in LA traffic. We had the tools already, so I said, "OK. Let's try this!" It's been over two years."
>
> David LaRosa, CEO and Creative Director, Verso

Fighting the need to work remotely will only cause your operations to flounder. You will waste two valuable resources: time and money. On top of that, you will put the morale and the outlook of your team in jeopardy.

Remote Working Is Older Than You May Think

To help relieve anxiety about a sudden switch to remote working, business leaders should realize the long history that exists in working from home with nontraditional or flexible hours. The idea of flextime was created in West Germany in 1967. German manufacturer Messerschmitt-Bölkow-Blohm developed flextime to combat work issues such as employees showing up late and poor morale, which resulted in low productivity.

Allowing employees to change the set of regular hours they worked but still work the same number of hours had incredibly positive effects on performance. It also had a dramatic effect on commuting. Having entire shifts of people all entering and exiting the building at same time created congestion and anxiety for workers to combat.

The term *telecommuting* was first coined in 1972 by engineer Jack Niles when he was interviewed while working remotely on a NASA communications system. The gas shortage of the late 1970s prompted author and economist Frank Schiff to push the idea for the American people to work from home. He wrote an article for the *Washington Post* suggesting that if only 10 percent of the millions of work commuters would do their jobs from home two days a week, they would greatly reduce pollution, road congestion, and employee stress levels. They would also reduce the long lines at the gas pumps where it was not uncommon for fights to break out due to the limited availability of gasoline.

Suffice it to say that flexible work hours and telecommuting have a long history. They were created to solve problems. It is far from a new concept, and it has been proven to be effective in boosting morale as well as productivity. But how many people actually had the opportunity to work remotely pre-Covid-19 (Figure 1.2)?

Americans Increasingly Working Remotely

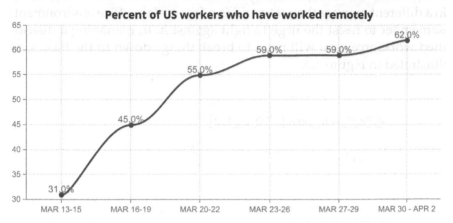

Percent of US workers who have worked remotely

Figure 1.2: Percent of workers who had ever worked remotely during the beginning of COVID-19–related closures.

Source: Modified from Megan Brenan, U.S. Workers Discovering Affinity for Remote Work, Gallup Pannel, 2020.

Productivity and Job Performance

Although remote working has a long history, a Gallup poll in early March 2020 showed that only 31 percent of workers polled had ever worked remotely. That means that at the time, management of remote workers was also not commonplace. Then on March 19, 2020, the World Health Organization declared COVID-19 a pandemic. In an attempt to reduce the spread of the disease, businesses that were not considered essential either shut down or had their workforce do their jobs remotely. This change caused a massive spike in remote work in a very short period of time, from one third of people polled having had any remote work experience to nearly half just a few days later. That number would increase to over 60 percent by the end of the month.

It is one thing to manage an occasional work-from-home employee, but suddenly switching your operation to be fully remote takes management adaptation. Part of that will be changing how you gauge productivity and performance of team members.

How managers measure productivity depends on the type of work being done and management styles. The remote workflow will proceed in a different manner than it would in a face-to-face office environment. Remember to resist the urge to fight against it. In managing a distributed workforce, you will need to break things down to the basics, as illustrated in Figure 1.3.

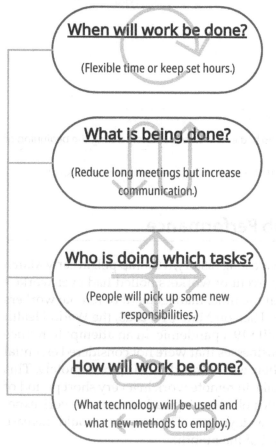

Figure 1.3: Immediate remote workflow elements

These will be your first key elements to look at when getting your remote operations started. Keep in mind that this will be new for the team as well. Make sure that your team has the tools they need not only to work but to thrive. Take care to choose the apps and services that are right for your team, such as software as a service (SaaS) providers with good customer support and an easy setup process. SaaS is a hosted subscription business model in which you pay a monthly or annual fee to use an online program and its features. Netflix is a great example of an SaaS provider. The subscriber pays a monthly fee to use Netflix software applications to watch shows and movies on demand.

You may pick some SaaS tools and find you want to switch to some different ones a few weeks later in the transition. That is perfectly fine. Many of the people interviewed in this book did switch some apps and services during their crisis in order to better serve their client base and facilitate better collaboration between team members. It is best to cast away the tools that don't fully serve your team than to carry on using them and letting frustration build. Don't be afraid to ask providers to allow your team to try services out before you commit to a contract if the software or service providers don't offer a standard trial period.

You will also need to consider that your employees may not all have the personal equipment your company needs for them to do their best work from home. They may have slow personal laptops or poor home Internet connections. These are obstacles that can be overcome but will need to be addressed. The company may have pay to alleviate those work-from-home issues in order to keep some employees and maintain productivity. At the end of the day, those expenses will be a great investment.

Adjusting to the New Work and Home Balance

In getting accustomed to your new work situation, there will be both excitement and anxiety. Humans are creatures of habit. Getting up, dressed, and ready for a commute to work is a routine that is difficult to change. A sudden break in that routine has an effect on people that many times is negative. Creating a routine for the day will become essential to your adjustment.

Also consider that, for some, work is a kind of escape from their personal life. If an employee has an unhappy home life, suddenly being at home all the time can be difficult for their morale and outlook. It is important to keep these situations in mind when measuring individual productivity during your switch.

Social Isolation

As a leader, it is important for you to look to increase your observation and listening skills when it comes time to get everyone working outside of the office. Some people may need a bit more attention than others. That in no way makes them a weaker link in your chain. Everyone adapts at different rates, and not everyone has the same lifestyle. Keep in mind that they are literally taking work home with them for the foreseeable future. Worlds are colliding. Their personal world and professional world are being mixed in a virtual blender, and it will take some adjustment for them and whomever they share their life with at home.

BEST PRACTICES

Your listening and observation skills will become more important when it comes to employees during the transition to remote work. Look for signs of anxiety and stress in individuals on your team.

Nicholas Bloom is an economist who is widely known for his work on the benefits of working from home. Bloom followed 1,000 workers from the Chinese company Ctrip who began working remotely and in 2015 published a report detailing his findings. After months, they saw productivity was up and that employee quit rates were down.

However, in an article in the *Stanford University News* in March 2020, Bloom reported some interesting additional findings:

> After nine months of allowing those employees to do their jobs at home, Ctrip asked the original volunteers whether they wanted to keep working remotely or return to the office. Half of them requested to return to the office, despite their average commute being 40 minutes each way.
>
> Why was that?
>
> "The answer is social company," Bloom says. "They reported feeling isolated, lonely, and depressed at home. So, I fear an extended period of working from home will not only kill office productivity but is building a mental health crisis. [5]

Employee Engagement

Team leaders and management will need to up their game when it comes to employee engagement during a crisis in which employees become distributed workers. Keep in mind that this will be a new and uncertain time for everyone. It is up to leadership to steer the ship in the right direction. Communicate proactively. You won't get those micro-moments that you do in the office when you run into people in the hall or turn to someone for a quick confirmation during a meeting when an idea strikes.

Check in regularly. A morning team check-in has become the norm for most companies that had to go fully remote. That is part of the daily schedule along with a host of other meetings for team member collaboration. At the same time, although it sounds contradictory, it's important to cut out unnecessary and long meetings. In future chapters, we will discuss where chats and apps such as Slack will come in handy in place of structured meetings.

Some people are accustomed to having the boss or peers just a few feet away. They prefer to have people around them that they can turn to in order to discuss an idea or even just crack a joke. The sudden change of being away from others in the workplace can have a profound impact on people. One realization that came out of the COVID-19 pandemic is that social isolation is a very real issue for many.

Some of the interviewees for this book have shown some great ingenuity in order to increase engagement and combat the isolation that comes from this sudden shift to remote work. From virtual happy hours to creating Zoom rooms just to sit with each other while they do work, the possibilities to create new virtual ways to connect and interact are nearly limitless.

I interviewed JP Holecka, CEO of Vancouver-based company POWER-SHiFTER Digital, for this chapter. The full interview is at the end of the chapter. JP had discussed with me the impact that suddenly going fully remote had on his team. To help combat the feelings of being isolated and to enhance online collaboration, he decided to have employees engage in daily standup meetings. He described them like this:

> Daily standups are where the project team spends 30 minutes discussing the project work they completed the day before, the project work that they have ahead of them for that day, and what they are intending to get accomplished by the end of the day, and lastly if there are any things blocking them from getting their things accomplished. After the standup, if there are any blockers, the people that are involved with that project will coordinate to figure out a way to solve the problem. It's borrowed from agile project management philosophies.

> *JP Holecka, CEO of POWERSHiFTER*

Daily and weekly meetings are all part of the adjustments to the new work reality. During the crisis that causes you to have your company go remote, workers will need to have more structure and balance between their personal and professional selves as well as continue to be able to socialize with coworkers.

Maintaining Growth and Profitability

For a company to maintain its profitability during this period, you will need to better enable your people to do their work well. Enabling them is not just a role for technology. During this crisis it will be vital for you to maintain your company culture. That is what will help get everyone through the rough patches and keep your company thriving.

Even if everyone cannot be at the office together, your company culture should still be preserved. It is not something that is attached to any physical space.

The company culture embodies the company values and corporate mission. It guides the way when people ask, "What should I do?" It is so important to the success of a company that many leaders see a strong company culture to be even more important than a company's business strategy. I dedicate a chapter in my book *The Artful Ask* to this topic. Here is a snippet from that chapter:

> *What exactly is a company, or corporate, culture, and why is it talked about so often in today's business environment?*
>
> *Company culture is made up of the shared values, beliefs and behaviors that define a company's vision. It helps drive engagement and social inter- action, and it shapes how management communicates with employees. This culture also helps define discretionary employee behavior. In other words—what they do when the boss is not around. Think of it as a set of family values outside of the employee handbook. The company culture guides employees' day-to-day tasks, how they make decisions in the work environment, and frames their interactions with other team members.*
>
> *One of the reasons company culture has been the topic of news and articles in recent years is that executives and business strategists are waking up to its importance. Technology has helped to emphasize that importance.*[6]

If you work to maintain your employee trust and your company culture, it won't matter if you are all working together in the same building or if everyone is scattered across the nation. The foundation of your company will still be there in the form of your culture, and that is where you truly work from. It will guide everyone in their day-to-day decision making as well as how they respond to issues during the transitional period.

I am a driving, focused manager: do the task, solve the problem, finish the plan. In this day and age (after going remote), I now better understand the softer side of leadership. It's important I listen to my team, be empathetic to the challenges they are facing during this difficult time, say and do the right things to unite them and build a culture of trust. Those soft skills are actually more important than the hard skills. When people trust the organization, leadership and each other, they produce better work.

Wendy O'Donovan Phillips, CEO, Big Buzz

Let's Get to Work

When disaster strikes and you find your place of work unusable, it will be highly stressful for all involved. You will want to be able to transition your team to doing their work remotely in a way that will not impact your clients or customer relations. You cannot plan for every contingency in business, but you can plan for this one. Adding a robust and adaptive remote working strategy into your continuity planning begins right now.

Armed with the right tools and the right attitude, you can set your company up for continued success during your crisis.

The company culture that has been a well-tested foundation for your business will continue to guide your decisions and help to guide you in how you respond to the crisis. The technology discussed in this book will merely be tools to further enable that culture and allow your people to keep doing their best for the clients and for each other.

It will be up to you and the leadership team to stay focused on your company mission and maintain a high level of engagement with your employees regardless of where the work is completed. You will best do this by creating avenues of better collaboration and communication. Doing this will also help you to maintain a culture of trust, which will be needed for everyone to meet and overcome the challenges that will arise.

You may also find some opportunities that come with your crisis event. One of the businesses profiled in this book leveraged its own crisis to see its company in a fresh light. It went from being a B2B company to a B2C company during the COVID-19 shutdown. It acted in accordance with its mission statement of helping people and stayed true to its corporate culture. By focusing on those things during the crisis, the company was able to see its organization from a different vantage point. The crisis helped reveal what the organization was really good at doing, and how it could best serve those that the company was created to help in the first place.

Another interviewee was able to use an existing service that their company offered in a new way to help clients keep their own businesses thriving during the shutdown. Innovation in business typically comes from recognizing a need. As Benjamin Franklin once said, "Out of adversity comes opportunity."

In short, a great amount of good can come from tumultuous times. If you have yourself properly prepped, you will find that you can continue to make meaningful connections with clients and co-workers remotely. You will find that your company can thrive during your tough times and come out stronger. That ideal is exemplified in the following interview with JP Holecka, who found his company suddenly working remotely twice within two years.

So, with that, let's get to work.

JP HOLECKA, CEO OF VANCOUVER POWERSHiFTER DIGITAL

Company Profile

- Location: Vancouver, BC
- Employees: 20
- Primary Line of Business: Digital product and design studio
- Primary Audience: We service product owners in enterprise-sized and scale-up-sized organizations that need to work with outside agencies to design and develop digital products and services.

About the Company

POWERSHiFTER is a digital product and service design studio that partners with leading brands and scaleups to produce digital products—applications, mobile apps, and websites—that unlock the greatest amount of value possible for your brand's digital touch-points.

For many companies, from startup to enterprise, it's hard to allocate the in-house time and talent and support the agile and iterative culture required to build the caliber of application needed. It's even harder to do it fast enough that a competitor doesn't get there first—or worse, do it better.

Our passionate team lives and breathes product design every single day; it's all we do and have been doing for more than 10 years. We have partnered with numerous companies and used design thinking processes to make the complex simple. Achieving simplicity for your digital products is never easy, but it wins the hearts, minds, and brand loyalty of your customers: 64 percent of customers are willing to pay more for your products and services when they are simple and easy to use, and 61 percent of customers are more likely to recommend your brand because it provides simpler experiences.

Summary of Primary Services Offering

We bring 11+ years of practice to help clients navigate the complexities of designing, building, and scaling the right service for their audience.

Enterprise clients work with us to augment their internal resources, taking advantage of our outside perspective and "drop-in" product expertise.

Scale-ups partner with us to validate their products and improve their user experience, making sure they're able to scale with growing needs.

Service Design Strategy and Design

Our service design experts start by talking directly to your end customers and team, conducting field studies or workshops that observe what people do—in context. After we uncover pain points and opportunities, we map them out so your entire leadership team can visualize the 360-degree perspective. The business roadmap is built collaboratively along the way because involvement early and often reduces risks and saves long-term costs.

Product Development

We build digital products that customers will pay more for and refer more often. The list of well-produced digital products your customers and employees are using is growing daily, and so are their expectations that yours will be as good as, or better than, the rest. A superior solution requires innovative and empathetic digital product design and development.

Web Design

We simplify websites to tell better brand stories and convert more often, whether starting fresh from the ground up or considering a long-overdue redesign. Does your site need to convey your brand story or convert visitors to customers? GI's primary service is providing full-service management for our association/nonprofit client organizations. Think about full-service management as the shared economy for our clients. Instead of each of our clients having their own office, staffing, office equipment, HR, finance staff, etc., we provide them with all of that, sharing those costs and resources among our clients.

How quickly did your company switch to working fully remotely?

We were ready for remote from a technical perspective. Everything we had from communications, collaboration, and storage was already in the cloud and every team member is issued a MacBook from day one. In 2018, we had to move into a new office that was under construction. We gave our notice at our old office and found out after our landlord had leased it that our new office would not be ready for over a month after we left our current space. That was our first get-remote moment, so when the [COVID-19] lockdown happened, we were more than ready on day one. That day was March 16, 2020.

How did working remotely affect your team working together?

Our team is an empathetic lot that loves to work with each other! We're very social inside and outside of office hours. So working together remotely was a challenge emotionally for everyone; some were really affected by the lockdowns. We worked hard to keep the team connected through daily all-company standups to start each day, and that helped a lot. We also invested in remote mental health programs that our team could access anytime for emotional and mental health needs. This was well received.

How did this affect your company culture?

The culture is strong in regard to being connected, empathetic, and supportive. The team has maintained all of the above very deeply, and the time invested in the years previous in that culture has really paid off. Our leadership team leaned in hard to make sure we were listening and actioning any of the new issues that might come up. I believe that also contributed to our culture being resilient. Our team really cares about one another and that matters in life.

What would you do differently?

I would have started our daily standups much sooner. We waited a month to move them from twice a week to daily. You really can get a read on everyone when you see them each and every day as a team, even virtually.

What advice would you give to a company that finds themselves suddenly working fully remotely?

Listen to your team, survey monthly, and action any of the challenges noted. Support them in new and innovative ways that they want, not that management wants. Our team did not want to do a lot of things that other agencies were posting about. Costume days on Zoom, forced games, or shlocky team events driven top down were not what would work with us. Your team will let you know how they want to adapt if you give them the opportunity, environment, and tools to do so. Go above and beyond as a leader.

Notes

1. https://assets.regus.com/pdfs/iwg-workplace-survey/iwg-workplace-survey-2019.pdf

2. https://www.wsj.com/articles/ibm-a-pioneer-of-remote-work-calls-workers-back-to-the-office-1495108802

3. https://www.forbes.com/sites/jennagoudreau/2013/02/25/
   https://www.forbes.com/sites/jennagoudreau/2013/02/25/back-to-the-stone-age-new-yahoo-ceo-marissa-mayer-bans-working-from-home/?sh=509148b01667

4. https://www.weforum.org/agenda/2020/06/coronavirus-covid19-remote-working-office-employees-employers/

5. https://news.stanford.edu/2020/03/30/productivity-pitfalls-working-home-age-covid-19/

6. *The Artful Ask: How Arts Organizations Can Build Better Partnerships and Lifelong Sponsors* (Pimbleberry, 2019), page 75

The Remote Workspace: Set Up Your Mind and Your Space

There was a time many years ago when I worked from home for about 10 months. Our company was still in startup mode, and the lease on the office space that we rented came up for renewal. We thought it a more strategic move to cut costs and work from home while we were engineering some new products and services.

I was able to set things up the way we thought we would need for my new home office to be the center of operations for a few weeks or months. I had a business-class SDSL line run to my house. I had two Cisco IP phones set up at a couple of workstations, some whiteboards, and a space for team meetings. We had the technology to do it seamlessly, so we gave it a shot.

I discovered that there are many different traps that can stop a person from being productive at home. From that comfy couch in front of the television to pets wanting attention, along with well-meaning family and friends, there are many distractions to lure you away from work. I myself ended up not being in the right frame of mind for weeks at a time on some occasions and found myself fiddling with home projects during work hours. I also began to notice patterns during my day when I would have periods of lowered productivity. The good news is that these work-from-home pitfalls are avoidable if you know what to look out for, you can plan ahead and set yourself up for success.

Your Work Zone: Distinguished from the Rest of the House

Just as your day-to-day place of work is distinguished from the other parts of your life, the same should hold true for your work-from-home space. Your work area should be where you go to work daily, and ideally, that place is different from where you watch television, eat, or sleep.

Not everyone lives in a place where you can have a separate and private space just for working. This may mean using a common area that is shared with the rest of the household. Do your best to choose a space where you can be most productive with few distractions. Heading daily to a bad space will have a negative effect on your mood, which will then be reflected in the quality of your work. It will also become apparent during video meetings. Picking the right space in your house or apartment will be crucial to higher performance, productivity, and your overall happiness. Choose wisely.

Choosing Your Space

There are some key considerations for choosing a workspace that will be shared with the rest of the home. First, try to find a space that can be dedicated to your work. Ideally, this is not a space where you have to pack up at the end of the day and unpack each morning when you sit down to work.

The dining room table is one example of a space that's not ideal. It gives you plenty of "desk" space, but unless you live alone, you can't commandeer the table just for your professional needs. You will have to put away your work items at least once each day at mealtime. If you can't find a space where you can keep your office all the time, don't fret. Use the time that you would have spent commuting back and forth to work to set up your work space in the morning and break it down in the afternoon. It will help give you a good sense of consistency and be part of a daily work routine.

Good Lighting

Do your best to find a place with good lighting. A space with plenty of natural lighting is optimal. Natural lighting helps to elevate your mood and can have a positive impact on your health and sleeping patterns. Do your best to position yourself and your monitor so that you avoid shadows and glares.

If natural light is not an option, be sure to have adequate lighting in the work area. Over a prolonged period of time, poor lighting will wear on you. It causes eye strain, which can lead to headaches, both of which will dampen your productivity.

BEST PRACTICES

If at all possible, pick a spot to set up your work zone that has plenty of natural light. It will help keep your mood elevated and reduce eye strain.

Broadband

Pick a location where you will have a strong Internet connection. In most homes, there will be spots where the Wi-Fi connection is weak or even dead. Avoid those spots because they will give you nothing but frustration.

I have helped design thousands of Wi-Fi deployments of all sizes across the United States. Many times, signal issues can be improved with some easy adjustments. There are plenty of ways to improve your Wi-Fi signal at home. For starters, you can alleviate those dead spots by raising the Wi-Fi router up higher and removing metal objects from its surroundings.

Wi-Fi is a line-of-sight technology. If you can see the router, you will have the best connection. However, every time the signal has to go through a wall or a piece of furniture to reach your laptop, the signal becomes weaker. If your Wi-Fi router is surrounded by clutter, or tucked behind a flat-screen TV, your signal will be far from optimal. Clear away objects from around the router and get it up to a level where it sits above furnishings such as couches and shelving.

Metal reflects and blocks Wi-Fi signals entirely. Therefore, if there is metal between you and the router, your signal strength and reliability will be affected. Metal racks and pipes can also interfere with the signal going back between your devices and the router. Some older homes and apartment buildings have chicken wire in the walls. These structures act as a Faraday cage and trap wireless signals. You may think that having spaces in the metal rack will let signals through, but it will also have a scattering effect.

Water also has a detrimental effect on wireless signals. If you have an aquarium, a fishbowl, or even plants between your workspace and the wireless routers, you will have some data issues. It sounds surprising, but the water will refract the radio waves and can cause lower data transfer rates as the signal changes moving through the liquid.

Signal interference from other devices also play a role in lowered Wi-Fi speeds, similar to having a conversation with someone in a crowded room at a party. There are other voices, some of them louder, that your ears are also picking up. That can cause you to miss some parts of your conversation. That can also happen with packets of data going back and forth between your laptop and the Wi-Fi router.

Use your phone, tablet, or computer to search for available Wi-Fi networks. You will see the network names of connections that are not from your own router. Depending on your device, you may be able see what channel these other network devices are using. You may need to download an app such as Wi-Fi Analyzer if your phone does not have a setting to show you the radio channels of the Wi-Fi networks in range.

In North America, most Wi-Fi devices have a channel range of 1 to 11, but the bulk of them are shipped out set to the default of channel 6. Many people do not bother to change this setting, so you end up with a good number of Wi-Fi networks all using the same channel. As you can imagine, that channel gets crowded quickly. Log in to your Wi-Fi router and choose a channel not in use by other nearby networks and farther from the channels they may be on. For example, if you see many on channels 1, 6 and 11, move your router to channel 3 or 9. You will need the username and password for your router to make these changes.

If your wireless router has external antennas, you can purchase larger antennas that will help strengthen the signal. You can also purchase a Wi-Fi signal extender to get a better connection. These "boosters" are inexpensive and easy to set up. They can save you a great deal of aggravation from dropped connections and help speed up your work.

If your Wi-Fi network is open and does not require a password to use it, then change that right away. You do not want client information or sensitive work data being sent over an unsecured network. Using the password protection option on your router will stop others from getting onto your network and looking at files and other info on your connected devices. It will also encrypt the data that you are sending back and forth as you work.

If you have multiple dead spots that you want to cover, consider upgrading to a mesh network. It deploys quickly and is very easy to set up.

Monitors

Ensure that you have your computer monitor positioned well and at the best height for you. The best positioning for your monitor is directly in front of you. You should not be twisting your neck or turning your body to see what is on the screen.

Set the monitor in a place where it will avoid glare. You may also need to adjust the curtains or window blinds at different points during the day as the sun moves. Keep the monitor about an arm's length away from your eyes. You should be able to see the full screen without needing to turn your head.

Grab a Seat

Get a good chair. This will be the best advice for anyone who suffers from back pain. Sitting for several hours in your dining room chair just won't cut it. They were not designed for that purpose. In fact, I have worked in several restaurants that used chairs that were purposely uncomfortable to sit in after a while. This was done so that people would not take up the tables for too long after they had eaten their meals to allow for more frequent table turnovers.

Opt for a chair with lumbar support and a good cushion. You will also want to be sure it is at a good height for the table or desk where you will work. Remember that you do not know how long you may be working remotely. If you do not have a good chair to work from for several hours a day, invest in one. Even if you purchase a secondhand chair from a thrift store, you will be better off than if you have something that may lead to back or neck pain or even poor circulation. Figure 2.1 lists qualities of a good desk chair.

Invest in Yourself

I cannot stress this part enough. You are worth a few extra dollars to get some items that will make your work-at-home experience easier and healthier. Consider getting yourself a comfortable keyboard. I find that the best ones for me have pads to support my wrists. Shop for a comfy and full-sized wireless mouse that best suits the type of work you will be doing from home.

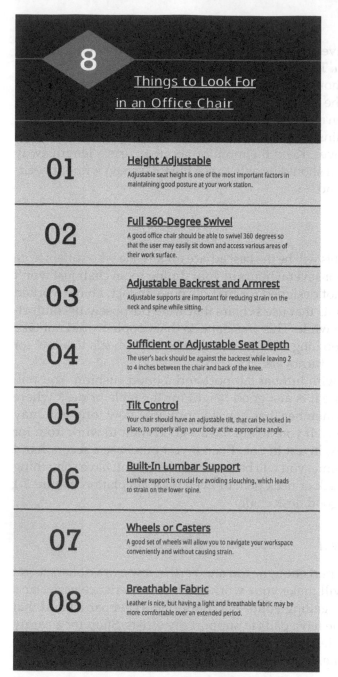

Figure 2.1: Eight things to look for in an office chair

Source: Based on John Triano, DC, PhD., Office Chair Back Support, Veritas Health, 2003. https://www. spine-health.com/wellness/ergonomics/office-chair-back-support

Also consider purchasing a UPS surge protector. UPS stands for *uninterruptible power supply*. In the US, you can purchase a simple one for under $50. Your laptop has a built-in battery, so should the power go out, you should not have an issue with it suddenly powering off. However, if the power goes out, your Internet modem will not work, so you will have no Wi-Fi. No Wi-Fi typically means that you won't be able to work until it is back up or you'll need to head somewhere else. If you connect the Internet modem and your Wi-Fi router to the UPS surge protector, then you will have Internet access for as long as the UPS has a charge. This will give you the time you need to wrap up a meeting and save any work you may have been doing online.

If you choose to invest in a better UPS surge protector, you can keep working online for several hours if you lose power. UPS surge protectors are rated by watts and volt-amps (VA). The lower the VA is, the less time you will have to power your modem and router. For most people, their Internet modem from the Internet service provider will also be the Wi-Fi router. If you have a separate Wi-Fi router, then you will have two devices plugged into the UPS surge protector, and it won't last as long. At the end of the day, a UPS surge protector is a sophisticated battery. The more devices plugged into it and operating, the faster the power will drain. The UPS surge protector will have a set of outlets dedicated for surge protection only and a set for both power and surge protection. Use the outlet set for power for the modem and router only so that you can optimize how long the battery will last to keep you working. Keep in mind that regardless of the size of the UPS surge protector you purchase, it will take several hours to charge. So if you do lose power, you won't be able to run to a store to buy one and then expect it work as it should. It is a solution for emergencies, but it's a solution that you need to plan ahead to use.

Add Privacy If Needed

If the household is too chaotic for the work you have to do, then create the quiet place you need by sectioning it off from the rest of the home. This means closing doors that may normally remain open to allow housemates or family to wander freely in and out of the area that is now your workspace. If you have a room divider, use it. A curtain rod and curtain in an entryway will also add privacy quickly, easily, and economically.

Sometimes, wearing headphones can help create the privacy that you need. It puts you in your own space and signals to others that you are not available.

It is not ill-mannered to create a wall of privacy if you need it. In the long run, it will be better for your relationships with family or roommates. It creates a buffer between your distinct work area and the rest of the living space. You are claiming this territory as your own, but be sure to not be too stringent in keeping people out during work hours. Remember, this is still a home to you and to them as well.

Kids will interrupt, pets will wander in, deliveries will come to the door. That's OK. Expect all of that to happen now and again. Keep up that understanding and you will be well poised to maintain lower stress levels at home while you work. You will also keep stress levels down and avoid misunderstandings if you set up some healthy boundaries.

Set Healthy Boundaries with Family and Friends

If you are working from home and you live with anyone else, you have to be cognizant that they will suddenly have a new roommate: the work version of you. That's an entirely different person in most cases. The work version may not be accustomed to housemates watching television at a high volume, dogs barking, cats jumping on keyboards, or kids playing raucously a couple of rooms over. Instead of letting these things become potential sources of stress, be proactive and have a talk with the household.

Your chat won't do much with the cat or dog, but explain the new situation to the rest and ask them to help you make this a good experience. Try to keep regular work hours and let them know when you will be available and when you won't. This will help them and you schedule work life and home life appropriately.

It's also a good idea to have a similar talk with family and friends who live outside your household.

No Is a Powerful Asset

You should do your best to learn how to say no during your remote work experience when it is appropriate. Since you are home, friends and family may invite you to socialize during your normal work hours. They may be innocently thinking that you will be home anyway, so there is no harm in going out for a few hours. It may be more fun to go shopping with your spouse or take the kids to the park or hang out with friends than it is to work.

You should for sure take frequent small breaks, but only you will know if a long lunch with a certain friend may turn into an all-day event. Let them know that you are not free during your set hours and reschedule for another day. It's OK to say no to others, and at times, you will have to say no to yourself as well.

Self-Motivation from Home Can Be Perilous

Being able to say no is going to be one of your best assets when it comes to staying self-motivated. Saying no to yourself will happen more often than you may realize at first. You are working from home, after all. Home is the place where you have fun and work on personal projects. This is a great argument to keep regular and scheduled work hours when working remotely. It allows for better segregation of the responsibilities of work and home.

Productive Procrastination

It's far too appealing an idea to catch up on laundry or get a head start on cleaning the kitchen in the middle of the workday. They are things that need to get done and dishes and clothes can just be thrown into machine to be washed. You will justify doing chores by the fact that you are still keeping yourself busy and doing something that is productive.

This is called *productive procrastination*, and it's a trap. Once you are in the home mental mode, it's hard to switch back into the work zone and be productive. Avoid doing chores during your work hours. Don't even do them on your breaks. Save the break times for a higher use, which we will discuss in more detail later in this chapter.

WORK NOTES

Productive procrastination: keeping yourself busy doing something other than what you should be doing.

Work and Play

Your home is also where you watch shows, play video games, and spend time on social media. Checking your social media or catching up on the news during a planned break or lunch is what you may normally do when at the office. Do those things, but be aware of how often you check them and how long you spend on them. You will be shocked at how quickly these short intervals add up to hours of time by the end of the week.

Keep yourself limited to those times just as if you were at your traditional place of work. If those things would not be appropriate at work, don't do them during office hours when working from home. Once you start down that path, it's far too easy to keep getting distracted by the call of entertainment. Put them out of mind.

Your Mental Space

When you set up a new work space, one of the first things you are likely to do is to tidy up that area. You take away any extra items that may add to the clutter of your work tools. You do this to be better organized and to eliminate distractions. In the same way, it's also important to de-clutter your mental space if you are to stay focused in your new work space.

Do you have a large number of open tabs on your browser? Close any of them that you do not need for the workday. If you have some personal blogs or websites that send you notifications when there are new posts, mute those notifications. Decide what is useful and what is not.

If you are a multitasker, consider switching to a single task methodology for a while. Multitasking keeps your mind full of minute details about the different tasks you are on. That adds to the mental clutter quickly. To keep focus and mental clarity, do one task a time, do it well, and then move on to the next.

You want to limit the amount of extra things that will steer your thoughts to topics that would take you off on a tangent. If you find yourself getting up to do something that you would normally do at home during your weekends, then it may be time for a short work break. Scheduled well, these breaks can help to keep you in the correct frame of mind and on the right track.

Take Breaks and Don't Feel Guilty

Hopefully you don't work in a sweatshop where work is nonstop with few, if any, breaks. That type of work environment is unhealthy and miserable. It leads very quickly to burnout, which makes the entire methodology counterproductive. Instead of peaking at a certain point of the morning followed by a downward spiral, you want to stay productive through the whole of the workday. To achieve that goal, make sure to give yourself time for breaks and do it without guilt.

Many of us have a mindset that taking a break is equivalent to goofing off. That outlook is part of our hard work culture. It's as if we have been trained to think that work is good and play is bad. That's not true at all. With children, playtime is good in that it creates a healthy atmosphere that relieves stress and clears the mind. It gives the mind a rest while also nurturing creativity. This holds true for adults as well. You need the break to get some rejuvenation.

You Are Not a Machine

Think of the time you spend dedicated on a work task as a type of gym workout. We will use the treadmill as our example. You cannot stay running on a treadmill for eight hours a day. After a certain point, you will go from a run to a jog to a walk and eventually end up in a heap on the floor. You need to take breaks. You need recovery time. A study published in the *Journal of Applied Psychology* shows that some types of breaks will serve you better than others when it comes to preventing work fatigue.

This study, "Give Me a Better Break: Choosing Workday Break Activities to Maximize Resource Recovery" by Emily M. Hunter and Cindy Wu,[1] demonstrates that what you do on breaks as well as when you take them have a dramatic effect on how fast you recover. In other words, there are better ways to take a break than what you may already be doing. You can also build a strategy to getting more out of your breaks by utilizing the following best practices.

Replenish Your Resources

It's helpful to think about the need to take good breaks in terms of resources. In my office, we use the term *bandwidth*. If asked to take on an extra project, I may respond that with everything else that I have going on, I just don't have the bandwidth to tackle something else. At any given time, you have only so much of any given resource that you can allocate for work responsibilities. If you use up your resources and don't replenish them, you will quickly feel them get spread too thin, and everything will suffer.

At work, these resources are *concentration, energy*, and *motivation*. You slowly deplete them as the day goes on. Breaks will help you recover those limited resources. If you are strategic about your breaks, you can slow down the depletion and even increase your recovery rate.

Take Breaks Early

The most important best practice tip is to take breaks earlier in the day. "We found that when more hours had elapsed since the beginning of the work shift, fewer resources and more symptoms of poor health were reported after a break," explains Hunter and Wu.

BEST PRACTICES

"Breaks taken early in the shift are associated with more resources after the break than breaks taken later in the shift."

Emily M. Hunter and Cindy Wu. "Give Me a Better Break: Choosing Workday Break Activities to Maximize Resource Recovery."

If you don't take your break until the afternoon, large amounts of resources will be already used up. It is harder to recoup those resources after they have been consumed down to a certain point. You will end up still feeling drained after that late-day break.

After analyzing data from close to 1,000 work breaks, their study finds that taking a break midmorning is far more beneficial for fighting workday fatigue. You will feel much more refreshed and recharged from a morning break than from a break later in the day. *When* you take your break is important, but so is *how* you take that break.

Take Better Breaks

How you take a break has an impressive impact on your ability to recharge. It is commonly believed that doing non-work-related things on your break makes for a better break time in terms of benefits. Evidence shows that this is not the case.

What you do on your break, whether work-related or not, is not the big concern. To take a better break, it is important that you actually enjoy what you are doing. Hunter says, "Finding something on your break that you prefer to do—something that's not given to you or assigned to you—are the kinds of activities that are going to make your breaks much more restful, provide better recovery, and help you come back to work stronger."

These better break activities can help combat fatigue, both mentally and physically. Stepping away from your desk and doing something enjoyable can help alleviate musculoskeletal symptoms and other types of work fatigue. Those who participated in the study reported that they experienced fewer symptoms of eyestrain, back pain, and headaches after an enjoyable break.

WORK NOTES

Personal Resources

Invest in the time to replenish your cache of personal resources: concentration, energy, and motivation.

Although the study did not narrow down a specific length of time for a break, it showed that frequent shorter breaks allow you to keep more of your personal resources. Taking a long lunch is nice, but having groups of short breaks is best when it comes to recovery. "Unlike your cell phone, which popular wisdom tells us should be depleted to zero percent before you charge it fully to 100 percent, people instead need to charge more frequently throughout the day," Hunter said.

Think of these breaks as an investment in your resources and in yourself. Take preventative breaks to make sure you are not dragging in the afternoon. It's a smart work strategy and it's good self-care. Make planning and taking better breaks part of your daily routine.

Create a Morning Routine Different from When You Were Commuting

To fully acclimate to working from home, you will need to create a new routine. You had one when you had to commute to work each day. You may not have thought of it as a routine, but the sequence of actions you took each day before you entered the building for work was your routine. You are accustomed to having that routine, and it's more important to your productivity than you may think. A mindful morning routine that you intentionally create can be a powerful tool.

Benjamin Spall is co-author of the book *My Morning Routine: How Successful People Start Every Day Inspired* (Penguin, 2018). The book discusses the importance of the morning routine from the perspective of

60 highly successful people who were interviewed. Here is an excerpt from the book:

The way you spend your morning has an outsized effect on the rest of your day. The choices we make during the first hour or so of our morning determines whether we have productivity and peace of mind for the rest of the day, or whether it will clobber us over the head. Unfortunately for most of us, good days don't happen by accident. Unforeseen events will step forward to challenge your best-laid plans. If you don't dip into your inner reservoirs of energy, focus, and calm first thing, you won't stand a chance. Start your morning with intentionality, and you can then bring these "wins" with you into the rest of your day.

Highly successful entrepreneurs give credit to their morning ritual or morning routine for helping to boost their productivity and outlook for the day. It drives them into the best mental space to get the day started and further fuels their progress throughout the day. From Oprah to Steve Jobs to Richard Branson, successful people have touted the morning as a key part of maintaining productivity and reaching your goals.

The need for a healthy morning routine is nothing new. Benjamin Franklin was famous for getting up at 5:00 a.m. each day. His daily routine is not far off from that of many successful people today: Get up and get washed up, meditate or pray, set daily goals, read, then eat. That routine set him up for a positive and productive day physically, mentally, and emotionally. Benjamin Franklin found it to be a recipe for success: He was not only a founding father of the United States, he was also a well-established entrepreneur and small business owner.

Set Daily Goals

Working from home means you are more isolated than you are in your traditional work environment. That is one great reason why you should set daily goals. Being alone all day makes it easy to lose focus. When that happens, you lose productivity. Daily goal setting is a way to combat that eventuality. It keeps your mind in work-mode and helps you stay focused on professional aspirations.

Tasks vs. Goals

For goal setting to be most effective on your productivity, it is important to understand the difference between tasks and goals. Think of a goal as a desired outcome that you can achieve. Tasks are actionable items. They are work to be done, such as items on a to-do list. Tasks should be used as tools that are aligned to help you achieve your goals.

To set effective daily goals, first look at what your long-term goals may be. Once you have a good idea of what those longer-term ambitions are, the daily goals will come to you. The daily goals are stepping-stones to achieve your longer-term goals.

Longer-Term Goals

Let's say you'd like to be a better leader, and you set that as a long-term goal. Communication is key to being an effective leader, so you may include improving your communication skills as a daily goal. Your task list may include proactively contacting a client on a potential marketing advantage that you feel they could leverage. Or it may include reaching out to a team member who you know is having difficulty with the isolation of working remotely.

Do those tasks, and you are accomplishing your daily goal, which set you up to achieving your long-term goal. Although they may seem small, accomplishing those tasks represents victories. Celebrate your victories. Reward yourself when you accomplish a goal. Give yourself a bit of extra free time, get yourself your favorite coffee drink. Do something. Achieving and enjoying the small victories helps keep your head in the game and maintains a positive outlook from your feelings of accomplishment.

SMART Goals

Just as there are better ways to take breaks, there are better ways to set goals. Consider using the SMART method for effective goal setting. In 1981, George Doran published a paper entitled "There's a S.M.A.R.T. Way to Write Management's Goals and Objectives." Since then, his method has become one of the most well-known and effective goal-setting techniques. Doran used an acronym in naming his methodology:

"WORK NOTES" AND "BEST PRACTICES" DO

SMART goal criteria are Specific, Measurable, Achievable, Relevant and Time-bound.

Specificity is key in any endeavor. The more ambiguous a goal is, the easier it is to be steered in the wrong direction.

Measurable results are important to be able to quantify your actions. If you cannot measure your results in a tangible way, it is difficult to see if what you are doing is making headway to achieving your goal.

You want to keep your goals **attainable**, otherwise you are simply wasting time doing busy work. It's a form of self-sabotage. Do you remember discussing productive procrastination?

The goals should be **relevant** to what you ultimately want to achieve. Look at what your short-term and long-term goals may be to judge the relevancy.

Last, you give yourself a **time** limit or due date to achieve these things. Goals that can constantly be put off will be put off. Procrastination is a trap that most people are familiar with falling into.

BEST PRACTICES

Set daily SMART goals for yourself and the team. Make sure they are specific, measurable, achievable, relevant, and time-bound.

Team Goals

If you manage a team, take team goals into consideration. Look at your desired outcomes and use them as your guide. What are team goals that you can help facilitate? Again, use the SMART goals technique to decide.

Let's use the example of a team goal of increasing traffic to a certain client's website by 30 percent before the end of the current quarter. This goal represents a SMART goal: it's specific, measurable, achievable, relevant, and time-bound. There are daily goals that can be set for each team member along with tasks that have an impact on this specific team goal.

This team goal can also be in sync with the individual goals of your team members. As a team leader, you can guide them in developing stronger search engine optimization or adding more appealing photos or graphics that cause more interactions with the client's website. This makes your team stronger and would help with your long-term goal of increasing your leadership abilities.

Remember that tasks can be important steps to success, but achieving the end goal is the prize. If you are feeling frustrated with something during your remote workday, knock out some goal-oriented tasks on your list. You will feel that sense of achievement and the positivity that comes from accomplishment and getting closer to your goals. Focusing on these daily goals will help keep up your remote working productivity and maintain your long-term visions that you have for your career.

Communication Habits: Respond Quickly and Use the Right Tools

Communication is going to be very different in the distributed working scenario. There is something positive to be said about an in-person conversation. There are cues that you pick up on that you may not be aware of until you are unable to have face-to-face conversations with people. When discussing a project with someone who is standing in front of you, you pick up on their body language, their intonation, and other nonverbal methods of communication. There are things that people convey with their eyes, body, and voice that do not translate well via a video call. This can cause a lack of communication or, more accurately, stunted communications.

Communicate Proactively

Your saving grace will be learning to be proactive in your communications while working remotely. Stronger communications will help you stay on the same page with everyone from management to peers on your team. When working in person, it's easy to pop your head into someone's office and have a quick conversation to clarify a point on a project. If you are a team leader, be sure to check in regularly with team members to ask how project flows are going. There are things that will get lost in translation from in-person meetings to virtual meetings. Expect that and compensate for it.

Set Expectations Early On

Be clear and concise in your expectations. Everyone who works with you on the team will be part of the transition to working remotely. You cannot assume they will all have the same expectations. If you are leading the team, then you are the one who has to set those expectations. Get together and figure out common work hours when people will be available for communicating with each other. Then discuss methods of communication. Not everything has to be discussed in a video conference or a phone call.

The Right Conversation Tools

I have some people on my team who work best when we go back and forth on Google Chat for a discussion. Others prefer a quick call because a verbal discussion allows them to best express their ideas and what they are trying to relate to me. Everyone will have to find what works best for them. Do your best not to force everyone to communicate one-on-one the same way as others. Not everyone expresses themselves the best in the same way as their peers.

There are Zoom rooms for more one-on-one conversations when the entire team does not need to be involved. Chat apps like Google Chat for G Suite users, Microsoft Teams, and Slack are popular for quick conversations. Find the right tools that work best for quick conversations and the right tools for more in-depth discussions.

Be Precise

You will find that you have to have a higher level of detail in your remote discussions. Again, much gets lost in translation remotely. There are nuances that simply do not get conveyed when a discussion does not happen in person. The level of detail that you use in your speech when conveying important points will become more important than ever when working remotely. Check your phrasing. Be precise in your language use. It may mean the difference between one meeting or several in order to make sure that everyone is on the same page. Part of being a strong collaborator and a good communicator is using precise phrasing.

Daily Check-Ins

The daily check-in will help set the priorities for the day and help avoid the urge to micromanage. Not being able to see everyone in person will cause a bit of a struggle in keeping everyone on the same page. It is easier to keep everyone on track when you are able to pop your head into someone's office to check in on progress. The daily check-in takes the place of that and allows the team to keep running smoothly.

This is another form of proactive communication. It keeps everyone up-to date and allows an open forum for people to ask questions, get clarification, and see each other every day. Being apart from each other is going to be tough on everyone. The daily check-in will become a welcome bit of normalcy that will become anticipated. It provides structure and regularity to everyone's workday.

I interviewed Wendy O'Donovan Phillips, CEO of Denver-based company Big Buzz. We discussed some of the hurdles of the team not seeing each other in person during the COVID-19 shutdown. I also asked her about some of the proactive communication methods she used during her remote work experience. Here is an excerpt from that discussion.

> We had to get very clear about who we are and why we do what we do, and really focus on the people within the agency. And I believe that when we focus on the people within the agency, they can genuinely focus on the people who are our clients, and the projects. We get the work done, but we have to have the relationship first. This has worked really well for us. We have used a once-daily morning stand-up meeting via Zoom where everybody goes around the room and says first of all personal good news, professional good news, what they're working on today, and where they need support. We need to do that sort of thing, because in the office we would share that just naturally. And I don't get that unless I connect with people.

> *Wendy O'Donovan Phillips, CEO of Denver*

Leave some time in the beginning of the daily check-in meeting for some socializing and catching up. It will help with morale and assist in maintaining your company culture. Helping to combat the social isolation that employees experience will be part of leadership responsibilities in the scenario where everyone is suddenly working remotely. Be ready to step up to the plate to address that need.

Make Yourself Available

Whether you are a team leader or not, you should make sure that people know when you are available. Colleagues should not feel as if they may be disturbing you if they need to ask you a question or raise a concern.

I go out of my way to let others on my team know what hours I am working and on which days. I also allow my team to see my calendar so they know when I have meetings set up or will be away from my desk. This gives them plenty of notice if they need to reach me. It's important for me as head of the company to be accessible if I am to be an effective leader.

If you are available, then set your status to *available* or *online* in chats and other communication apps that the team is using. It allows others to know that it's OK to reach out to socialize or share something with you. It facilitates better overall remote communications and allows the workflow to move along at a good pace.

Get Dressed for Work

Many bloggers came forward during the beginning of the COVID-19 shutdowns to give advice on how to dress when working from home. The advice from many of them was to dress comfortably. The argument was that if you felt cozier and more comfortable, it would help to reduce stress that was abundant from the health crisis.

That time at the beginning of the shutdowns was filled with plenty of uncertainty on professional and personal levels and the need to reduce stress was universal. However, at that time, everyone was dealing with that same crisis. Because of that, people were more forgiving about how we acted and dressed for remote work. Huge portions of the population were suddenly doing their jobs remotely. But the next time most companies face a crisis where they cannot work from the office, the situation will not be so universal.

Dress the Part

The cliché "dress for success" is widely known, and there is truth to it. How you carry yourself and your physical appearance have a direct effect on how others perceive you. This will be true if you are meeting in person or on a video conference. Decide what kind of impression you want to make in your professional life. Remember that, and then dress accordingly for your remote workday.

> **BEST PRACTICES**
>
> Dress for work. Business casual is appropriate for most audiences. Business formal will project your role as an expert in your field.

Business Casual

For most cases, full business attire is not needed when working remotely. Far from wearing sweats or leggings, the business casual attire is a good option. This has several advantages. For starters, it can be a good part of your daily routine that sets the tone for the day. Wearing sweats does not give you a mental preparedness of productivity, unless you plan on cleaning out the garage. But when you get dressed as if you are heading to the office and you know you look good, you can feel it too.

You can get a mental confidence boost by how you dress. Many have experienced getting dressed up in new clothes to go to a party and feeling great as they walk out the door. It's the same in the work environment, regardless if it's at the office or remotely. It sets your mind up for success. It also sets how you interact with others.

Know Your Audience

According to Qualified Communications, in April 2020, 30 million people were using Zoom meetings to conduct business, whereas only 10 million were doing the same in December 2019. Qualified Communications conducted a survey of remote workers to see what their experiences were on video.

The survey of 465 men and women[2] of various ages were asked about the impression a speaker's appearance made on them with regard to certain professional traits. The specific traits asked about were trustworthiness, authenticity, expertise, and innovativeness. Regardless of age or gender, people surveyed found that business casual attire gave the speaker the impression of having three of those four traits (Figure 2.2). The one exception was expertise.

Perception and Impact of Clothing Formality

AUTHENTIC

- Casual   26.4%
- Business casual 46.4%
- Business formal 27.2%

INNOVATIVE

- Casual   17.6%
- Business casual 52.6%
- Business formal 29.8%

EXPERT

- Casual   7.3%
- Business casual 26.0%
- Business formal 66.7%

TRUSTWORTHY

- Casual   18.7%
- Business casual 48.6%
- Business formal 32.8%

Figure 2.2: The perception and impact of clothing formality

Source: Based on Quantified communications virtual meeting survey March-April 2020. Quantified Communications, Inc.

In most cases, business casual is your best bet for remote work if you want to project authenticity and trustworthiness. If you are going to be meeting with new clients remotely or presenting to senior executives, you may want to project yourself as an expert. If so, then wearing more formal business attire is the better choice.

Think of how you get dressed for your remote workday as putting your best foot forward. For the most part, dress the way you would if you were going to the office. It sets the tone for your perception of the day and how you are perceived as well.

Set Yourself Up for Success Right from the Start

Finding yourself and your team suddenly working remotely is going to cause stress and anxiety, and you will have a learning curve. Your first instinct is to rush right in and get everyone going. I would agree with that course of action, but experience says to take the time to first get your gear set up correctly before you dive fully into these new waters.

Getting things set up the way you will need them will reduce your anxiety and increase your confidence during this experience. Speed is of the essence because you need to keep the workflow moving and the company operational, but be sure to move in a calculated fashion with an intentional methodology. First keep in mind your duties, the team, and the clients, then move forward with your setup from there.

Taking better breaks will keep you in the right mental space and help you replenish your personal resources. Communicating proactively will save time, keep everyone on the same page, and reduce the feelings of isolation while helping to maintain morale. These are self-management and team management tips that will be highly useful during this switch to remote work.

Your daily goals and daily routines will help you stay on track and motivated to meet the day while keeping you consistent throughout the week. Getting dressed appropriately will help you stay in that professional state of mind while you project authenticity to your team and your clients. The following interview shows how this business leader took seriously the importance of presenting himself professionally for virtual meetings with clients and his team.

Of course, all of these remote work capabilities are made relatively simple to achieve today, far more so than even 10 years ago. The technology that is available to you is powerful and convenient. Choosing the correct technology for you and your company is paramount to your success. We dive deeply into that topic in the next chapter.

MATT SMITH, CEO
SPEAK2 SOFTWARE, INC.

Company Profile

- Location: Morganville, NJ 07751
- Employees: 5
- Primary Line of Business: Health and wellness
- Primary Audience: Adults ages 30–60 who have aging or impaired loved ones

Brief Company History

We were founded in April 2019. We formed based on the need we saw for aging adults to stay connected and have a way to communicate with the people around them in a way that made sense. Screen-based technology has become the dominant form of communication, and it has shut out many people who are aging. Our company was formed to change that.

Summary of Primary Offering

Speak2 is a platform that connects aging adults to their friends and family by allowing them to communicate in the easiest way possible—using their voice. Seniors can send and receive messages, record activities, and make requests for assistance through an Amazon Echo (Alexa). The circle of trusted friends and family can interact using the Web, app, or voice devices. We connect the screen generation to our aging loved ones.

How quickly did you go fully remote? What was the trigger that day you went remote so suddenly?

We pretty much did it right away. New York and New Jersey were hit hard and early. Our office in Newark closed, and we all had some personal connection to the virus—family members. I myself had it in late February before it was known. So we were very quick.

What was the most difficult part of going remote for you and the team? (Feel free to discuss going from B2B to B2C, if you'd like.)

The most difficult part internally was the disruption to the routine and having to spend time learning new ways to collaborate. But for us, the disruption was actually because our business is working with senior living, which basically went into quarantine and lockdown. So our clients couldn't engage with us.

They became inundated with calls, paperwork, and many other items related to COVID, so working with us became a very low priority. They literally had people dying in their care. While our service was clearly helpful from a communication perspective, it just didn't get the attention of the people working in or running these communities.

The positive part of it was it forced us to examine our product and make it better, make it more useful and easier to adopt. We ended up shifting much of our business from senior living communities, like assisted living or nursing homes, to *anyone* who is aging, wherever they are.

We became a platform for those that are "aging in place." So our actual business model expanded from B2B to B2B and B2C. This is a major shift and we are still rolling things out, but we are very lucky we did because it has gotten incredible results.

What was the easiest part of going remote for you and your team?

I already had a home office, so it was more a matter of making it an everyday office instead of a once-in-a-while office.

What part of your management style needed to be tweaked with a remote team?

Communication needed to be more concise and clear. In person, there is some body language and immediacy of clarifying as someone gets into work. They can just pop over and ask a quick question to clarify a topic. When working at home, it requires the Zoom call, so it's not immediate and time can be wasted until that next video call is set up. We use Slack quite often, but on the more strategic, deeper topics, being in person has some advantages.

Maybe the biggest change was not having an easy way to do whiteboard sessions. We were evolving and changing our product. Being in a room with people and being able to visually whiteboard as a team is really important, and we had to make do with video calls instead. We got better at it, and I ended up using a Bamboo tablet to mimic whiteboarding, but it was not the same.

How did you set up your workspace at home? Tell me how you set it up and the investment you made in that space.

I invested in lights and hung them on my walls. I did research so I have three overhead lightboxes at three different angles and two LED lights on the floor facing behind me to eliminate shadows.

I made my office more "background friendly" by putting my daughter's artwork on the wall. She is 16, so these aren't children's paintings, these are *really* good. And I have my guitar on a nice stand to give some character and show some personality.

I also have a projector screen that I can pull down and give myself a pure white background if the situation calls for it. I invested in a Stream Deck so I can more easily navigate my screen using single button pushes during demos.

I already had a Yeti mic but I purchased a boom stand since it was being used so much.

Does your team use any unique or key processes in your day-to-day work?

We have a daily scrum meeting to cover product and marketing. We use Trello and HubSpot to manage our tasks and customer interaction.

How did you alter these processes to work remotely? What apps, services, or technology did you use to bridge the gap from in-person to remote in order to keep these processes alive?

Based on being remote, we use Slack and Zoom as our regular way to meet. We try to mimic being in the office together by making sure we are looking at each other for all of our meetings.

We all invested in some lighting and high-end cameras as well so that the quality is as good as it can be. We find that it is not a good experience, and does not encourage collaboration, when the lighting is bad, the camera is blurry, or if someone looks like they are in a messy room, etc.

The environment matters. While we understand that people may be balancing work/home craziness, something simple like two good LED lights makes all the difference in the world. We've also encouraged our people to get condenser microphones, such as a Yeti, which is what I use. We reimburse our employees for these items, and it's well worth it.

What would you do differently if you had to suddenly switch your organization to working remotely or if you could do it all over again

If I could do it all over again, I am not sure I would change much. I'm really proud of how we handled things and that we used it as an opportunity to evolve instead of an excuse to give up on things. If anything, we've come out the other side in a better place. We adapted quickly, used tools to help us adjust, and got to work. Maybe I would wear a mask sooner.

What advice would you give? What words of wisdom and advice could you give to someone else who finds themselves having to suddenly make their operations go remote? Are there any pitfalls that people should look out for?

My advice would be to quickly research what is available to help. Take an inventory of what you will be missing and determine if it's a good or bad thing. Some of the changes might be a blessing in disguise or present an opportunity to alter things that you wanted to change anyway but were too complacent to do.

Probably the biggest piece of advice I would give is to quickly get the tools needed and do not take shortcuts. The microphone on your laptop or home computer is not going to be as effective as an external microphone. Getting a hi-definition camera and lighting makes a huge difference. The few hundred dollars, if you can afford it, is well worth the investment to create a professional and engaging work environment from home.

Notes

1. https://doi.apa.org/doiLanding?doi=10.1037%2Fapl0000045
2. https://www.quantifiedcommunications.com/en/resources/report/video-communication-trends

3

Office Technology: Stay Connected and Competitive

Transitioning the entire team from the office to remote work from home can be relatively painless if you have the right tools readily available. As previously discussed, with continuity planning, you should ideally have a plan in place prior to your crisis. It's a preventative strategy to allow you to recover quickly with fewer interruptions to your business during a disaster.

It is best to migrate your current operations now to using the methods and technologies discussed in this chapter so that recovery from your crisis point will be a quick one. If you prep your day-to-day work operations for remote transitions now, you could stay up and running and switch the team to full remote work the very same day a crisis occurs. If done in this manner, clients may not even become aware of the transition.

VoIP: From Desktop to Soft Phones and Features

Back in 2004, I co-founded a B2B company that provided business communications services including broadband and Voice over Internet Protocol (VoIP; sometimes called *IP telephony*). VoIP allows users to make and receive phone calls over broadband instead of using plain old telephone service (POTS), in other words, traditional telephone lines.

The business class landline services from the big telecom companies have come a long way in recent years and offer great features for small businesses, many of which can now be managed from a centralized online call management portal. If your office becomes unusable, you could use the online portal to forward those incoming calls to another phone number. If your company does not receive a large number of calls each day, this could work as a good solution for your day-to-day business needs and emergency remote work needs. However, if the business uses the phones quite a bit as an external communication tool for sales and customer support, then it's probably best to look to a VoIP system.

There are very few disadvantages to using a VoIP system for your business. The only major one would be that that you need a reliable broadband connection with good upload and download speeds. Your office should already be equipped with a sufficient broadband package from your local cable provider or the local telecom. So let's take a moment to look at the residential speeds required for remote workers to use VoIP at their homes.

It is recommended that you have at least a 3Mbps connection both for upload and download for a stable quality of service. You can get away with less to make calls; however, you need to take into consideration other services that may be using the same Internet connection. If you are trying to make a VoIP call at home and someone else is streaming a movie while someone else is playing a video game online, you may drop packets. When that happens, your voice may get choppy to the person on the other end of the call, and you may miss some of the conversation while experiencing packet loss.

A few years ago, this may have been more of a concern. However, broadband speeds for residential service have sped up considerably and continue to increase. With this in mind, there is virtually no reason to not make VoIP a part of your remote work crisis prevention plans. It is a feature-rich service that will help make your transition simple and fast.

BEST PRACTICES

Switching to a feature-rich VoIP service for your work operations will ensure that you will not lose any calls in the event of a crisis.

Cloud Management

VoIP systems are hosted and managed by the provider and use a software as a service (SaaS) business model. One great advantage to SaaS is that the phone service will work anywhere as long as you have a stable Internet connection available for your calls. Even if the Internet connection at the office is offline, many of the features of your VoIP phone system will remain active and clients may not realize that you are out of the office. Plus, you won't lose any calls if disaster strikes.

You will have a dashboard that can be accessed by administrators from any Internet connection. This is perfect for scenarios such as those we discussed in Chapter 1, "You Can't Go to the Office: Where Do You Go from Here?," in which an office may become inaccessible. It will be a useful tool for both the beginning and end of your company crisis. You can login to the dashboard, make the switches needed that best suit your remote work migration and then switch them back once everyone is back in the office. It will be a seamless and stress-free way to keep the phones ringing to support sales and clients.

Automated Attendant

The automated attendant will be the first interaction with your company when a call comes in. This is the outgoing message that a person hears when they call your number. A prerecorded message greets the caller and gives them the options to dial extensions for individuals or departments. This feature is hosted in the cloud, so it is not dependent on the physical accessibility of your place of work.

This automated attendant will also add to the seamlessness of sales prospects and clients reaching your distributed workforce. In fact, paired with the other VoIP features, there would be little need to change the outgoing message to let anyone even be aware that your team is now working remotely.

Forwarding and Follow Me

Cloud management over individual extensions will help keep the flow of business steady. Forwarding is fairly simple to explain. If a client were to try to contact you via phone at your extension, you can simply forward that call to ring your mobile phone instead of your desk phone. Again, this is a seamless transition due to the cloud-based dashboard.

The dashboard is set up to send incoming calls for your extension based on device. Typically, the device would be the phone at your desk, but you can set it up to ring at another desk phone should you want a co-worker to answer your line. Or it can be set to ring to another phone altogether. It is a device-based feature rather than a location-based feature. The dashboard essentially points the call at the device that you prefer to ring when a call comes through for you.

Forwarding also allows you to manage voicemails. Normally when you have an unchecked voicemail, your office phone will flash to alert you to check your messages. However, you can forward the individual extension to email the voicemail instead. This allows the recipient to listen to their voicemails as an audio file attached to an email without ever having to be in physical contact with their desk phone.

Follow Me is slightly different. Let's say that you are expecting a highly urgent call and want to be absolutely sure that you pick up the phone when it comes in, regardless of where you may be at the time. You can program the dashboard with a series of phone numbers where you may be, and the call will be pointed to find you at any of these phones.

For example, the first phone to ring may be your desk phone. If you do not pick up, the system will then send the call to your mobile phone. If you still do not pick up, then the call would go to your home line. The system will try to reach you at each phone number that you have programmed, in the order you designate. Or, it can be set to ring all these different phones simultaneously. The call will thus follow you, allowing you to stay connected to your clients regardless of where you may be.

Soft Phones and Mobile Phones

Should your team not be able to take their VoIP desk phone with them to work remotely from home, you have other options to keep them in contact. You may already be familiar with a soft phone. A soft phone is software that sits on your computer or tablet. You will see a virtual number pad that allows you to punch in the outgoing telephone number; then, you can use a headset or the speakers and microphone on the computer or tablet for the call.

The interface is designed specifically to be highly intuitive and resembles the typical office phone handset. It is complete with a number pad, a call screen to show caller ID, a hold button, a transfer button, and more. The soft phone is also an economical way of getting started with a new VoIP system since you would not incur any equipment costs for physical desk phones.

You may already be familiar with soft phones due to the fact that Google offered a now defunct soft phone service with Google Talk. Skype also has a popular service for consumers to allow people to make local and international calls over a computer or tablet. Skype also provides a business version of its soft phone service. This business-class service is called Skype Connect, and it can integrate with a company's existing VoIP system.

Many VoIP service providers also offer a mobile phone app in their arsenal of business call solutions. Once installed, this would allow team members to make and receive calls from their own mobile device as if they were using their desk phones. The app would give the added benefit of allowing the team members to retain the privacy of their personal mobile phone numbers while conducting business for their company.

Due to these features, a VoIP phone system is a more robust option over POTS lines for a small business when it comes to the ease of remote working. It empowers your team to keep working and brings a sense of normalcy to your crisis. One thing that you do not want to worry about is being able to answer the phone when it rings. Being able to reliably answer sales and support calls will help keep operations seem like business as usual as far as clients are concerned.

Email, Chat, and Channels

There is a good deal of discussion about the customer experience and the brand experience. These are examples of how consumers interact with a company. Marketing professionals take their work regarding the brand experience seriously, and with good reason. Creating positive feelings or even emotional connections to a brand will help build faithful and sometimes lifelong customers. Because of this, the ways companies communicate externally to their customers are well measured and given a great deal of thought in order to maintain a great, consistent brand experience.

How a company communicates internally has a big impact on the employee experience. That is something that should be taken as seriously as the customer experience. Savvy businesses leaders know the importance of how a company communicates with staff, especially when it comes to attracting and retaining talented employees. A company that overshares information in a deluge of redundant communications can get a reputation for stifling creativity and productivity with needless amounts of communication.

When everyone suddenly begins to work remotely, there will be some communication adjustments due to the simple fact that you can't just walk over to a person and have a face-to-face conversation. Using the right technology for sharing various types of information will help keep the workflow at a good pace and maintain a positive employee experience

BEST PRACTICES

Create a company communication policy. It should cover how you communicate internally and externally. Be sure to review what medium is appropriate for each type of communication and collaboration.

Email

For most companies, the biggest use of emails is for external communications. Email is a great tool for marketing campaigns as well as detailed communications with clients and vendors. However, for internal communications, email is far from the best way to convey quick bits of information. In fact, many large companies discourage the use of internal emails. The main reason is that excessive emails kill productivity.

In the modern workplace, people receive so many emails that not all of them are even opened, much less read. Most people have hundreds of unopened emails in their in-box. Because of this digital deluge, the attention span for reading emails has been drastically shortened, much to the chagrin of many email marketers.

For internal use, emails are still appropriate for company memos, policy changes, or announcements. For quick conversations, though, it's best to use chats.

Chats

Have you ever been in on an email chain that seemed to never end because people kept adding in small bits of input? These email threads can cause a great deal of frustration. They are confusing and they make it difficult to get good information in a timely manner. Plus, they can kill enthusiasm for a project. Chats help prevent such issues.

Google has had Google Hangouts as part of its Gmail suite of business services since 2013. It was a part of Google's messaging service and offered video chat as well. In October 2020, the tech giant introduced

Google Workspace, a centralized suite of applications for conducting business communications. This includes Google Chat, which takes the place of the familiar Google Hangouts chat.

Since my companies have used the Google suite of services for years, I am very familiar with the chat service. It's an ideal way for me to have a quick conversation with one person to get and give info for a project or client. It's simple to send a link or an image on the platform and is similar to using text messages on my phone. Just as with text messaging, I can have group chats with multiple people in real time. This is a great solution for my teams because we already use corporate Gmail services, but there are many other chat services available to non-Gmail users.

One of the most widely used chat platforms is Slack. In fact, many of the business leaders interviewed for this book use Slack for their companies, even if other collaboration tools that they use have a similar messaging feature. But even this great communication app can be abused and misused.

Remember the never-ending email thread? Imagine being in an ongoing group chat where everyone chimes in randomly throughout the day. The conversation can devolve quickly and become more of an ongoing ramble. Worse yet, when it gets misused in this way, it can make people less responsive.

If someone was in a meeting or away from their keyboard for a bit, they can quickly find themselves awash in a sea of chat messages discussing a variety of topics. They try jumping in, but they have no context and no idea of the direction the discussion is going. This kind of misuse can lead to a productivity nightmare. Use of channels can be a preventative measure to keeping discussions on track and uncluttered.

Channels

Channels is a feature available on Slack as well as on other collaboration tools that feature chat apps, such as Microsoft Teams. In short, they are a way to keep the conversation organized by topic or project. Even a specific channel can have multiple side channels for different subtopics or areas of work. This keeps information organized and stops overlapping discussions.

BEST PRACTICES

Leverage channels to keep conversations organized and productive.

For example, let's say you have a channel dedicated to a specific new client. That channel is where the whole team gets together to discuss the project, but the folks in marketing can have another channel just for them to discuss their tasks dedicated to the new client. At the same time, the team from product development can have their own channel for that client and sales and distribution can have their own channel to discuss their roles and action plan. This allows each team to coordinate within their own field of responsibilities as well as with the entire group. It keeps the conversations cohesive and timely.

Channels can also be dedicated to socialize with each other, share good news about the company or individuals, or just have fun. Remember that during your crisis, there will be loads of stress and feelings of isolation. The teams will need a way to stay in touch and share things with each other that are not entirely work related. They would do that if they were in contact with each other in the office, so it should be encouraged during your crisis as well. It's part of a healthy employee experience, just as it is for them to sit and talk together at the office. There is a way for them to get close to doing that from home as well.

Voice Chat

Discord is a service that has primarily been used by gamers. It allows people to create a private server, similar to a channel, and chat in real time using VoIP, text, and video technologies. It is great for people who want to team up and play an online game together and speak live in real time as they play. It's also a nice tool for developers who may be working on a project together and want to work live. It's even useful for a group that just wants to sit down and shoot the breeze.

One of the great features of Discord is that its channels can have up to 250,000 members and support 25,000 simultaneous online members. The applications for Discord on a business use level still have not been tapped, but for social discussions, it is already a heavy hitter. For virtual face-to-face connections, the video meeting apps are your best bet.

Videoconferencing

Due to the inability to meet in person during the COVID pandemic, the popularity of video calls and videoconferencing exploded. One of the most popular platforms for video meetings is Zoom. It is a great example of the proliferation of video for business. In December 2019, Zoom reported 10 million daily meeting participants using it service. In April 2020, it reported 300 million daily meeting participants on its platform.

Zoom is one of the biggest players in the industry, but there are many different videoconferencing options available. Choosing the best one for your company will depend on your specific needs, the size of your company, and any integrations with third-party apps that you may want to use.

During the pandemic, many businesses began using video in ways in which they probably had not previously used it and would not have used it if they were still able to meet in person with clients and other people on their own team. That led to a great deal of innovation in how to use the technologies to conduct business, manage teams, socialize, and maintain revenue streams. How you will use it is entirely up to you, but you will probably find more ways to use video during your crisis event.

I interviewed Leslie Murphy, president and CEO of Raybourn Group International, and discussed how working remotely affected company processes. She told me how video meetings unexpectedly changed how her company will conduct internal meetings in the future. Here is what Leslie had to say:

> Videoconferencing has changed how we do things. A couple of our people were already remote. They weren't ever going to be here in this office. Sometimes if we had a team meeting, people would meet in the conference room and then would audio in the person who was remote. We're not going to do that anymore, because we found that when we're all on the same level playing field people engage much better.

> *Leslie Murphy, Raybourn Group International, Inc.*

Internal Meetings

Skype was used pre-pandemic by many smaller companies for instant messaging and quick meetings by people who worked remotely a few days a week. However, when the entire team goes remote, the need for more features and functionality became apparent. Figure 3.1 summarizes the top four videoconferencing solutions. We will discuss each below.

4

Top Services for

Videoconferencing

Videoconferencing has become a crucial tool in the modern workplace. There are many options available, all with their own unique features and applications.

 ## Google Meet

Google Meet is a great solution for users of Google's Workplace. With Google's strong focus on integration, meetings can be scheduled in your Google Calendar, they can then be recorded and saved directly into your Google Drive for later access. With over 2 billion active Google Workplace users, you can begin to see why this is such a compelling option for users and organizations who are already working in a Google environment.

 ## Microsoft Teams

Microsoft Teams is another powerful solution that has seen tremendous growth based heavily on its integrations with the Microsoft Office 365 Business platform. More than just videoconferencing, Teams has a strong focus on collaboration through features like document sharing, instant messaging, and screen sharing but with added security features for protecting sensitive information.

 ## Cisco Webex

Cisco Webex is an enterprise-class videoconferencing solution offered by Cisco Systems. As a world leader in the development and manufacture of network and telecommunications hardware and software, it comes as no surprise that you can expect a host of security features and a high-quality audio and video experience. Webex also features real-time transcription and closed captioning, making it a great option for the hearing impaired.

 ## Zoom

Zoom exploded in popularity during the COVID-19 pandemic and it is easy to see why. While they are not as focused on integrations with other office applications and environments as many of their competitors, what makes Zoom stand out is their ease of use, reliability, and accessibility. Zoom is a great option for one-on-one meetings but with support for up to 1,000 meeting participants.

In a survey of 4,700 users, the top benefits of videoconferencing were cited as:

94%	88%	87%	87%
Increased efficiency and productivity	Increased impact of discussions	Expedited decision-making	Reduced travel

Figure 3.1: Top four services for videoconferencing

Source: Adapted from The Business Benefits of Video Conferencing, Parmetech

Daily or weekly video check-ins with employees is crucial when your team is distributed in multiple locations. These check-ins keep everyone on track and create a forum for employee feedback. At the end of the day, video calls or conferencing are communications tools. They will take the place of an in-person formal meeting, the social get-together, and presentations. Look at your company needs and choose the tools that will best fit the size of your company. Be sure to take into account what other services and applications you are already using for your day-to-day operations. You want to be sure that services work well together or even complement each other.

External Meetings

For external meetings, I prefer to use video services that do not require the end user to download an app. I have been in a number of online video meetings that began late because one or more of the persons invited to the meeting had trouble installing the meeting app on their computer. Imagine the embarrassment of trying to close a sale to a prospective client and one of their people who is a decision maker was not able to attend because they couldn't get the app installed on their computer. Or worse, they got so frustrated that they chose not to participate in the meeting at all.

Even though a third-party videoconferencing app not controlled by your company caused the bad experience, it will still reflect poorly on you. Make sure that any app that you go with is simple to use if you will use it for external communications.

Google Meet

Google has a great video option for companies already using Google's suite of business platforms such as corporate Gmail and the Google Calendar. Google Meet syncs easily with its other business apps and makes scheduling and making changes simple. It's a reliable option for both internal and external communication needs. It's easy to use on both PC and mobile devices.

Cisco Webex

Webex was founded in 1995 and was acquired by Cisco in 2007. With that kind of longevity in the industry, Webex has learned how to offer high-quality video meetings. The service has exceptional audio clarity and offers real-time audio transcription as well as closed captioning. This makes it a great option for the hearing impaired. Being a Cisco company, it also offers a host of top-of-the line security features.

Another great feature of Webex is its virtual whiteboard. It allows you to share your screen as a blank whiteboard just as you would have on the wall in a brainstorming meeting at the office. Then you can use a variety of tools to draw what you need others to see. It also allows anyone or even *everyone* in the meeting to add to the whiteboard simultaneously. This is great tool to allow the free flow of ideas when you're not able to meet in person.

Microsoft Teams

If you have a larger organization and you are already using Microsoft Office 365, then Microsoft Teams is a superb option. It is fantastic platform for internal communications. Meetings can be scheduled in Outlook and easily added to participants' calendars. It's easy to share Excel, Word, or PowerPoint documents with team members. It also will transcribe meetings and make the transcripts available shortly after the meeting is completed. Microsoft also uses its AI to filter out background noise to ensure crisp voice quality during meetings.

Zoom

Zoom is so widely used that it is now part of pop culture. Zoom memes and jokes are all over the Internet and social media. Its massive popularity is mostly because its platform is simple, reliable, and feature rich. Many other video meeting platforms scramble to mimic Zoom features.

Security was a concern early on, but the company has added end-to-end encryption to help alleviate many issues. It integrates well with many third-party business applications. Zoom also offers integration with conference room hardware such as Polycom SoundStation and Cisco conference phones. That makes it a solid option for companies of every size.

Remo

One type of event that suffered heavily during the pandemic was conferences. Although it was still possible using videoconferencing technologies to see and hear a speaker offer a presentation, that is only a part of what makes up a conference. Conferences are a big part of revenue from the annual budget for many organizations.

General ticket sales only make up a fraction of the revenue for a conference. Event planners also work with sponsors of various levels, from table sponsors to title sponsors and many levels in between.

For the attendee, a big part of going to a conference is the networking. I enjoy speaking to people I see at booths at a trade show and mingling with people who are assigned to sit at company tables. Those are all elements that are missing from most virtual conferences: lost revenue and lost opportunities for the event host and attendees. Remo adds all of that back into the mix.

With Remo, attendees can be assigned to individual tables mapped out on a floor plan of the virtual event space. Attendees can then choose a table and ask to interact with the people "sat" there. This allows for a more natural flow of networking and even team-building activities. Sponsors can have banners placed throughout the event space and the platform offers options for them to collect emails and interact with attendees.

Remo touts itself as adding a more human connection for virtual meetings. It adds a fantastic level of interactivity and is a great solution for both fundraisers and event planners.

Team Collaboration Tools

I tend to see team collaboration and project management tools as separate but related necessities for remote teams. The developers with whom I work reside in different states in the US and in different countries around the world. Needless to say, I have been using group collaboration tools to help create new features for my own company's suite of services for years.

Collaboration tools had always been separate from project management tools. With online project management tools, once the project is developed, beta tested, tweaked, and finished, the project is put to bed. However, team collaboration for a company is an ongoing process. It should not end. Communication and collaborative relationships should continue to strengthen. Keeping that in mind, many of the tools we will discuss here started in one of these two areas but due to the pandemic changed their focus to add more elements of the other.

Team collaboration and project management collided into the same platforms. Some have showed real strengths on one side but weakness in the other. A company can't be everything to everyone. With that in mind, take a look at your team and your company needs. Some of the services discussed in this chapter would be best suited for companies that are heavy on the development side, such as website design or building apps. Other platforms discussed are better suited for companies that provide services such as bookkeeping and accounting.

Some of the weaknesses in these platforms may not be of any consequence to you, whereas some of the strengths may be game changers for your management requirements. You may not need an all-in-one solution and instead choose a project management tool that integrates with your existing team communication apps. The great thing is that you have many options to get your company to the right place when working remotely.

Trello

Trello launched in 2011 and is used by developers the world over. It uses a visually simple and uncluttered workspace called a Kanban board that allows for ease of use. Think of sticky notes on a project board. Those notes or cards are lined up in buckets on the board. For example, one bucket would be work that needs to be done, another bucket would be dedicated to work in progress, another might be for work ready to be reviewed, and one for work finalized. Tasks can be assigned so that everyone has ownership of each task.

This visually simple layout allows for you to log in and see everyone's progress at a glance. Trello also looks great on a mobile device. It is a very task-oriented platform that is highly intuitive to use. It offers templates designed for a wide variety of industries so that you can get a project started quickly and easily.

Basecamp

Basecamp began in 1999 and also offers a visually simple layout that allows you to see everything at a glance. It adds more communication tools for the team and keeps a record of responses from each team member, which is great for accountability. It's easy to see which tasks have been accomplished or are still pending.

Basecamp offers automatic follow-up alerts for tasks that are overdue. Project members can store files, schedule tasks, and keep track of an unlimited number of projects. Its simple design makes it easy to get started without much of a learning curve. Basecamp offers a flat fee for companies regardless of how many members use the service.

Asana

I consider Asana to be more of a work management platform than a project management one. The company was founded in 2008 by Facebook co-founder Dustin Moskovitz and former Google and Facebook product engineer Justin Rosenstein. Asana is a highly intuitive work tracking, task management, and collaboration tool.

It does feature the Kanban board as well for the ability to see where a project is going in a single look. But it also allows for different styles of project views, such as list or timeline view. This allows the end users to see the project in a way that is best suited to their priorities and personal preferences.

The free version allows up to 15 team users and includes basic task management features. The uncluttered design of this platform makes it a fairly straightforward tool to get a team onboarded and managing company tasks quickly.

Teamwork and Teamwork Desk

The Teamwork platform is a stable and reliable service with many add-on features for added fees. It has milestones that help you keep a keen view of where your projects should be going and what steps still need to be taken to get them across the finish line.

It has features for accountability and time tracking. The ability to track the time spent on each task is a great way to show clients who may want a more detailed understanding of the work that your team has accomplished.

The add-ons do cost extra but are worth a look. For example, Teamwork Desk creates some automation that allows your company to interact with clients and let them open trouble tickets. This is a great feature for companies that offer support services.

Slack

Slack is a communications tool for the workplace, but it is certainly not limited to any physical place of work. Imagine an instant messaging service on steroids. A team can use it to relay quick info and share files and links.

By organizing chat rooms, called channels, you can keep topics and team conversations from overlapping. A good example would be a creating a customer support channel. Sending a message to the support team and resolving customer issues becomes efficient via support channel chats.

Slack integrates with a large number of third-party apps, such as the customer relationship management tools (CRMs) Salesforce and HubSpot. This means that you can add in Slack for team communication and collaboration regardless of whether you use Microsoft Office or Gmail or most other popular business operations services.

Slack is one service for which you will want to set expectations on how it is to be used by the team. Getting inundated with constant messages can be distracting and become a productivity killer. Although Slack may have started out as mainly a messaging service and chat app, it is determined to stay a player in the work collaboration field. It has added in voice and video features along with productive tools to help maintain a healthy workflow.

Microsoft Teams

Microsoft Teams is a very robust and feature-rich collaboration platform. This is to be expected due to the Windows operating system and the pervasive use of the Microsoft Office suite of services. Its collaboration platform is often compared against both Slack and Zoom. These comparisons alone tell you that Microsoft Teams is a powerhouse for team and client communications.

When you take into consideration its integrations into Office 365, OneDrive, and SharePoint, then you quickly see that Microsoft Teams can be a centralized hub for team collaborations at every level.

It's a platform that is not just limited to Windows devices. You can use Microsoft Teams on Windows, iOS, macOS, and Android. It works great on laptops, tablets, and mobile phones.

Google Workspace

Over the years, Google has made a huge name for itself in the business productivity arena. Like Microsoft Teams, Google Workspace creates a single space for your company collaboration and communications needs. When used with its wildly popular web browser Chrome, Gmail, and its cloud storage service Google Drive, you have a scalable system for both remote workers and traditional employees.

Google Workspace simplifies getting on board with its service by offering only four levels of service. Companies that include only a single individual, enterprise-class organizations, and everything in between can find a price package suitable to take advantage of the technology. All levels of service include a rich suite of tools including Jamboard, their interactive collaborative digital whiteboard.

If your company needs more specialized services, they've got that covered with the Google Workspace Marketplace. Similar to Google Play, the marketplace is a one-stop shop for your business applications. Choose administrative apps, project management tools, CRMs, and many other categories. You are likely to find apps for services that your company is already using, so you can keep them on board as well if you choose Google Workspace.

File Sharing and Cloud Storage

One of the best ways you can stay working if the office suddenly is unusable is to proactively keep your documents centralized in the cloud. Most small businesses will not need the full services of an end-to-end encrypted cloud storage service. Such services are typically used for highly sensitive data and confidential files.

Some companies choose to have their data stored in house on a network attached storage (NAS) device, which is essentially a server connected to the company network that is dedicated to storage of company files and can be accessed by authorized people in the company.

The advantage of this would be the control of the hardware and the lack of any monthly or annual fees. However, many small businesses do not have an IT person to set up or maintain their own server. Another caveat is that should the server be inaccessible or the network goes down due to a disaster, you most likely would be unable to get to those assets for some time.

Having a third-party storage service that is off site can save you from worry since it is unlikely that there will be a scenario where the storage service would be unreachable.

When choosing a cloud storage service, it is important to understand the different services available, such as cloud storage, cloud sync, and cloud backup. Although they perform different but related functions, these services do not need to be used as independent solutions for your company. In fact, the services complement each other to help keep your digital assets safe and help increase workflow productivity.

Cloud Backup

Online backup, or cloud backup, gives your company a safety net should apps or files on one or more of your business computers become inaccessible. Think of it as a copy of your laptop. It houses all the customer files, company files, documents, photos, videos, and apps. If your laptop was lost or damaged, those assets on your hard drive may be lost.

You can even use cloud backup to restore all of your company laptops if need be. Should a laptop become damaged by a software virus, you could restore the laptop to a system image stored in your cloud backup. A system image is a full copy of your drive. It can be loaded on to the laptop once the virus is gone and fully restore your device back to the point it was at the last backup.

Cloud backup is insurance against lost digital assets. The strength of that insurance, however, is dependent on how often you back up your devices.

Over the years, I have heard too many horror stories of lost work on a client project or proposal all due to the infrequency of the backup. That's where automated backup comes in. You can set the cloud service to back everything up on one or more devices at scheduled times of the day multiple times a week. Some of the more respected and reliable companies in this industry are Barracuda, SolarWinds Backup, and Carbonite.

BEST PRACTICES

Protect your company's digital assets with cloud storage. In a disaster, this becomes a vital method for your company to access and share files to keep your workflow going uninterrupted.

Cloud Sync

Although cloud sync stores copies of your documents, its functionality goes beyond simple storage. Let's say, for example, that you were working on a client document at home on your phone or home computer. You make some changes and save the changes to the document and then head off to the office. Once you're there, your work laptop would already have the document stored but with all the changes that you made to it at home.

This feature is also a great productivity tool for team collaboration. As changes are made by various team members, each would have the most up-to-date version of the document available to them on their own devices. These services will sync across devices using Windows, MacOS, or Linux as well as other operating systems.

For small businesses, Dropbox has been a well-known leader in the cloud sync industry, as has SugarSync. These companies offer backup and cloud storage services as well.

Cloud Storage

The primary goal of cloud storage is to have all of your files in a centralized location. This allows any authorized users to be able to access those files at any time from anywhere. This is a perfect for operations with a physical office or remote employees or a combination.

Cloud storage allows a distributed workforce to be able to do their jobs and share files with freedom and flexibly. At the same times, the business also has the insurance of protecting its digital assets off site. This also frees up space on the hard drives of work devices to help to ensure that they operate at better efficiency.

Google Drive, Microsoft OneDrive, and IDrive are some of the top-rated companies in cloud storage. Always consider the other apps that you use for your company in order to choose who you will work with for storage. IDrive also touts a feature that helps protect companies against ransomware. That brings up the topic of security and what level of security you and your company would need in order to feel safe.

Data Protection

The cloud services discussed here all offer basic levels of security. The more you need, the more cost is involved. If you want to add an additional layer of security on your side, you can encrypt your files before you send them to the cloud. Companies such as AxCrypt and CertainSafe have software services that help keep files safer from hackers.

If your company works in the health, education, legal, or financial sectors, you may need to increase your security even more. Services certified by the Family Educational Rights and Privacy Act (FERPA), the Gramm-Leach-Bliley Act of 1999 (GLBA; also called the Financial Services Modernization Act of 1999), and the Health Insurance Portability and Accountability Act (HIPAA) are available from select cloud storage vendors such as SpiderOak, SmartVault, and Box.

Be aware that use of these certified cloud storage services does not make your company automatically HIPAA or GLBA compliant. That is a process for your own company to go through. These third-party services merely support your company's internal process in the endeavor of industry compliance. How you handle security on your end is always something that you should keep in mind.

Cybersecurity

Your company should have a cybersecurity policy in some form that should include general guidelines for working remotely. Even if it is not a formal written policy, you should have some security strategy in place to mitigate potential threats against breaches and losses. How you handle security when the team goes fully remote for a while should be taken into consideration regardless of whether your data is highly sensitive or not.

Wi-Fi

At the very least, ensure that employees are not using open Wi-Fi networks. When working remotely, make sure that they only use password-protected networks for work purposes and that their devices are not discoverable on public networks. Ask that they are also not using the default password that came with their Wi-Fi router at home.

Securing Devices

Work laptops, phones, tablets, and other devices should be password protected. Make sure that strong passwords are used with upper- and lowercase letters as well as several symbols and numbers. In fact, it is recommended that you use a passphrase instead of a password. Think of it as a short sentence, such as "I-L0v3_1ce_Cre@m!" Using a phrase is a bit more complex than a single word. A longer password better protects against brute force hack attempts. Passphrase protections will help ensure that only authorized people access company files and data.

Ensure that all work devices have antivirus and malware protection installed. Set them so that they update regularly. Update the operating software and apps on work devices consistently. Those updates typically have security updates along with them.

Two-Factor Authentication

Two-factor authentication (2FA) or multi-factor authentication (MFA) provides another layer of protection for accounts instead of just using a passphrase.

Using 2FA is recommended for email, your company social media accounts, cloud storage, company communication apps, and any company financial needs, such as online banking. Many online services such as Amazon, Instagram, and Intuit now ask you to enable a two-factor authentication for your accounts. If prompted to do it, you should take them up on the added security.

If you want to add 2FA on your own, there are apps for that as well. Companies such as Duo, Authy, and Okta all offer companies strong MFA solutions.

Remote Wipes

If a work laptop or phone is stolen, you may want to take steps to ensure that client or company data is not stolen as well. One solution is a remote wipe. This will erase all data on the device.

You may already be familiar with Apple's popular Find My Phone app. It features a data wiping ability should a user's iPhone become lost or stolen. Android also has a similar app, and you can add this feature onto laptops by loading the software from third-party providers. Prey is a good example, it is an open-source app that allows you to protect phones, laptops, or tablets from data theft. The service will help you track and recover devices that may be lost or stolen. It will also wipe those devices should they not be recovered.

Virtual Private Networks

You can add even more security by investing in a virtual private network (VPN) service, which lets employees access work applications via a secure encrypted network connection. It is an added operational expenditure, and some companies may not feel the need for its use because they may feel that their data is not very valuable to most hackers. On top of that, there are some CIOs who feel that such services slow down productivity in a remote work environment. Using a VPN is up to the business owner or IT manager.

Keep in mind that not all employees use company laptops when working remotely. It is difficult to enforce work security policies on an employee's personal equipment. You can mitigate risk that may come with employee personal behaviors and devices by using VPN services. A VPN will even help protect company data while the user is on public Wi-Fi networks.

NordVPN, Cisco AnyConnect, and OpenVPN are some of the top providers for business VPNs. When shopping for a VPN solution, you will want to look for reviews that mention speed, reliability, ease of use, and cost effectiveness.

Business Tech Advantages

As we discussed in this chapter, there are key technologies and services that will be valuable tools to help you and your team keep your company working as you go fully remote for a period of time. These are items that you should consider immediately to set yourself up in the best position for when disaster strikes and forces your team to work away from the office. Positioned strategically, these technologies can help streamline your operations and improve workflow in your everyday work environment and in your remote work environment.

During your crisis, when the phones start ringing at the beginning of each day, you want to feel confident that those calls will be answered. VoIP services and a cloud managed dashboard can help ensure that you will not lose any sales, support, or client calls.

Videoconferencing will allow you to maintain internal and external communications. From one-on-one discussions with team members to daily check-ins with the team to full presentations, this technology will keep the team on track with clients and with each other.

Many project management and collaboration apps do come with videoconferencing tools built in or as add-ons. But don't feel that you have to fully commit to any of them if they don't suit all of your needs. There is no reason to have to switch from a platform that you and your team enjoy using to take advantage of the benefits of other providers. These collaboration tools are set up to integrate with many of the leading business software and services available. That helps you get started with them and reduces your learning curve.

Creating centralized cloud-based repositories for your digital assets will not only protect them, it will also help keep your team working during a crisis event. Backing up and synching files can maintain productivity and insure against lost documents, records, and files.

You want to be aware of how those files are accessed by your remote team. They will be on different types of Internet connections with any manner of security setups. There are simply too many unknowns when the whole team goes fully remote. A bit of prevention now will save headaches later should you have a security breach. Be proactive with creating a cybersecurity policy for in-office operations and remote workers.

There will also be some fairly specialized tech and online services for each industry. Look to professional associations and groups geared toward your industry to see what solutions have been built that are customized for that industry. The following in-depth interview with Leslie Murphy from Raybourn Group International chapter discusses use of remote technology designed to help meeting planners and event professionals.

This chapter has identified some of the types of technology that your business should leverage to maintain operations and even thrive during your crisis. But how do you choose which software and which service providers will best suit your current and future needs? We delve into that discussion in detail in the next chapter.

LESLIE MURPHY, RAYBOURN GROUP INTERNATIONAL, INC.

Company Profile

- Location: Indianapolis, IN

- Employees: 32

- Primary Line of Business: Association Management

- Primary Audience: Trade and professional associations, individual membership societies, foundations, nonprofits, and other membership-based organizations

Brief Company History

Since 1988, trade and professional associations, individual membership societies, foundations, nonprofits, and other membership-based organizations have all trusted us as a leader that specializes in helping them thrive, not just survive. We build relationships with them, becoming a trusted and seamless extension of their team. Because of our personalized approach to expert association management, we become one with our clients, sharing in their passion and partnering in their success.

With more than 250 years of cumulative association management expertise, we serve state, regional, national, and international organizations with members in 107 countries.

Raybourn Group International is one of just 15 percent of association management companies (AMCs) accredited by the AMC Institute. Accreditation is the mark of excellence and it means we consistently demonstrate a commitment to uphold and deliver the highest level of customer service.

We are headquartered in Indianapolis, with service locations in the cities of Chicago and Memphis.

Summary of Primary Offering

RGI's primary service is providing full-service management for our association/nonprofit client organizations. Think about full-service management as the shared economy for our clients. Instead of each of our clients having their own office, staffing, office equipment, HR, and finance staff, etc., we provide them with all of that, sharing those costs and resources among our clients.

In addition, we also provide consulting and event management services for associations/nonprofits.

- Our consulting team is sought after to help organizations achieve positive results. Areas RGI has consulted in include strategy, membership, operations, governance structure, board orientation, and more.

- Our events team provides diverse services for clients executing in-person, virtual and hybrid events.

What was the reason you went remote so suddenly?

We have actually had part of our staff working remotely for over 20 years. The number has grown over the years as we worked to retain and attract great people to our team.

In early March, some of our Indianapolis-based staff started working from home as their children's schools were closing or they were concerned about their personal health. We also let our team know they had a choice to work remotely or in the office at that time. By mid-March, Indiana's governor issued a mandate that all offices should close and we did.

What apps, services, or technology did you use to bridge the gap from in-person to remote in order to keep the remote workflow alive with your team?

We recognized early during the pandemic that we needed to focus on ensuring that we were keeping our culture and supporting our clients.

Thankfully, we all had the ability to work remotely already. In addition to the staff that don't live here and work remotely all the time, many of us travel extensively and some work from home a day or two a week. So we focused on how to use technology to achieve our goals around culture and client support.

We went to Microsoft Teams to help with both. We had used Skype to IM with each other during the day, but Teams gave us new ways to communicate and engage for work and to support each other. As an example, moms on our staff started a channel in Teams to share challenges around working from home, suddenly homeschooling, or self-care. Good News and Celebration channels organically started to help combat all the bad news, share accomplishments, and just help people smile. Each of our client staff teams used the platform to connect quickly to find new solutions and engage more effectively than email. We used video even more to ensure we were really connecting with each of our team to encourage communication and also touch base with them on a personal level. For those not used to working from home, we wanted to ensure they felt very connected and supported.

Since Teams didn't allow all staff to see each other, we used Zoom for three new staff connections each week: Monday morning coffee, Whine and Lunch Wednesday, and Friday Happy Hour. These were important and voluntary ways the staff could connect at least once a week. Since we hired and onboarded five new staff virtually during this time, it was also a great way for everyone to feel connected and learn more about each other.

We had already gone to VoIP for our phone system with voicemails coming to our email, making changing where phone calls went and responding to clients efficiently. Heavy phone users added Zopier for great "real phone" functionality from their laptop.

What remote apps or services did you use for clients?

One of the things we do is to plan conferences for our clients. Since all of our conferences had to go virtual while still meeting the educational, networking, and financial goals of the organization, we needed to find a platform and quickly. We discovered Remo and utilized it for most of our virtual events in 2020.

The platform allows for presentations as well as for people to connect at tables or lounges and chat. While this could be accomplished in Zoom with breakout rooms, we found that Zoom didn't provide what the exhibitors and sponsors needed. Remo allowed for visibility for sponsors as well and good interaction for exhibitors at an affordable price. We also continued to use GoToWebinar for live and recorded webinars or a few of the early virtual conferences.

One last question: If you found somebody who, all of a sudden, has to work their company completely remotely, and they come to you for advice, what's the best [number one] advice that you'd give them?

I would say, remember your culture. Work intentionally on preserving your culture, connecting the people. Early on, especially. I'd say intentionally work on your culture because it will pay dividends when you get to the other side of this and jobs start to be pretty competitive again.

4

Choosing Technology: Getting the Right Digital Tools for You

Over the years, my company acquired many new clients who switched to our cloud-managed services after working with our competitors. Sometimes, these same clients were originally debating between our company and a competitor but decided to go with the other guy instead, for various reasons. I've made it a rule to ask why a company chooses to go in a different direction instead of working with my company after we've pitched them. Usually, the reasons come down to the price, promises, or performance that the competitor dangled in front of the client during the sales process.

When the other company overpromises on what it can deliver, it creates issues for its new clients. This causes frustrations to build, and those new clients quickly look for alternatives to get away from the vendor with whom they just signed on.

We tout our cloud-managed services as hassle-free. That's what a business should look for in its service providers. What I mean by "hassle-free" is a service that is reliable, needs little technical support, is easy to implement, and can scale to meet your company's needs today and a few years from today.

There are a huge number of options when it comes to business and office technologies. Throw in the increasingly competitive market of remote work software as well as collaboration services and you have an ocean of providers providing very similar yet different solutions.

This chapter helps you to navigate that sea of options. Here we focus on your current and future needs to guide you in making the best choices.

Scale to Meet Your Needs

One of the continuing goals for business owners is to grow their company, which involves increasing sales, obtaining new clients, adding more employees to the team, and maybe expanding the physical spaces to accommodate the growth. So that a business can do this without hitting too many growing pains, business plans typically include operational designs for scalability. Simply put, scalability is the capacity to perform well under circumstances of increased work. In other words, companies set themselves up from the beginning to handle increasing amounts of business. Designing company operations to be adaptive this way is intentional. Ideally, operations and methodology already allow for a good amount of scalability. To ensure success, the technology that a business uses should also be scalable.

In 2012, one of my companies took on a well-known new client for our cloud managed Wi-Fi. We would be adding our services to over 3,500 new locations in a span of about 120 days. We had been given a specific target date for when we needed to have our equipment installed in all locations. Normally, my company would manage such a system-wide rollout by coordinating our work with the other vendors on the projects, such as the broadband providers, network managers, and local installers. This helps us make sure that all needed systems are in place at each location prior to our equipment arriving to be plugged in and connected to the Internet. This also allows us to manage the speed of the rollout using our various management and monitoring systems so that we have the details of each location, such as the equipment placement within the building, what other equipment may be connected to ours, and names of onsite managers. These location-specific details are impor-tant to us for support purposes.

However, a new senior executive at the client's organization decided that he wanted to do things differently to make the individual locations less dependent on their headquarters. Instead of allowing us to work directly with the company headquarters to help centrally manage the timing of service installation, he wanted each individual location to control when they would order the services and have them contact us when they were ready. In other words, he wanted 3,500 locations to be installed by a certain date, and within a window of 120 days, but to leave it up to each location to reach out to us and place individual orders for our service at a time of their own choosing.

Unfortunately, people tend to be procrastinators. They put off some tasks on their to-do list to focus on more immediate priorities. If something does not need to be addressed for another 90 to 110 days, it will probably be put on the back burner until a deadline is imminent, at which time the project becomes a priority.

We pushed back on executing the provisioning of our service in this manner and mentioned the pitfalls of having 3,500 different decision makers in the mix. Our objections were heard, but the executive decision still stood. At the time, our internal systems and processes were already set up to work with clients that had hundreds or even thousands of locations, but we would work with a single point of contact at the main office for ordering new service and setting up recurring invoicing. For this project, we needed to set up something internally that was entirely different. It had to allow for thousands of location managers to place orders at any time and notify us so we would process the order, get equipment ready for programming and then shipping, check with the other vendors to see if they had their portions of work at those locations completed, and also have the location set up for recurring billing in our accounting system. And we had to get this new system up *quickly*.

We did our research, we spoke with some software vendors and put together an internal operational process and a customer-facing online system that we thought would work for the locations to place orders. At first it all worked really well. The orders for new service from the locations came trickling in at a good pace. At the end of the first 30 days, however, we discovered that the new system was not up to the task. The client and the individual locations were never aware of any hiccups, but internally we had issues and stress.

Some information from online forms was not carried into other databases, even though the databases worked during testing, the software plug-ins did not synch properly and were glitching at sporadic times. Loads of information ended up needing to then be manually entered by our team. At this time, only a few hundred locations had reached out to us, so we knew that there were a large number of orders still to come in for processing, which would likely cause a surge. The new system wasn't up to the promises made by our software vendors. We now knew it would not scale up to meet our needs, even though the software vendors kept promising that they could get it working the way it was supposed to operate.

We ended up finding a new software company that was able to help us accomplish our goals. The new processes we put in place also gave us the opportunity to take on a new type of client. But having the wrong Software as a Service (SaaS) system in place for the first part of the project caused us a great deal of frustration and caused numerous extra work hours for our team.

The reason that we were able to roll with the punches in these circumstances and still get the job done was that our main systems were designed from the beginning to adapt to a large increase in business. They were highly scalable. To keep your business successful in a crisis and allow for growth, the systems you choose to operate your company need to be scalable as well. Let's take a look at key factors in scalability as they pertain to your business.

Reliability

Many companies use online customer relationship management (CRM) tools that are far more than a glorified Rolodex to store customer contact information. The CRM system acts as a centralized database of your business's history with each customer. It houses communications that various team members have had with a customer, sales data, information on the customer's company, information on individuals within the company, and so on. It's a data-driven system used to build stronger relationships with the customers, which then builds sales. In many cases, it is also used for support purposes and marketing.

Companies who use a CRM system find it a crucial tool for their day-to-day business operations. So what happens if their CRM system goes down and nobody in the company can access client information? In most cases, the answer is that they have to wait for it to come back online. That means that the normal workflow for the day will be dramatically slowed down until the CRM system is usable again. That equates to lost time and money.

The Five Nines

There is a huge dependence on business technologies in order to keep a company working. That's why it's important to meticulously choose not only your tech but also who provides that technology for your company.

When a chief information officer (CIO) looks at new tech service providers, one of the first things they will look at is the service-level agreement (SLA). The SLA is a guarantee from the service provider that states what services they are providing with particular emphasis on the quality and availability.

The gold standard for a service provider's uptime is 99.999%. To many people, that seems like more nines than are necessary. However, the difference between an SLA of 99.9% and one of 99.999% is about nine hours of service availability over the course of a calendar year (Figure 4.1). That's nearly nine hours of downtime when the service will not be usable to the business that subscribes to it. If a CRM tool is down for nine hours in any given year, it's not as big of an issue as if the downtime involved a company's e-commerce website. Nine hours of downtime potentially translates to a great deal of lost sales.

The Time Difference of a Few Decimal Points

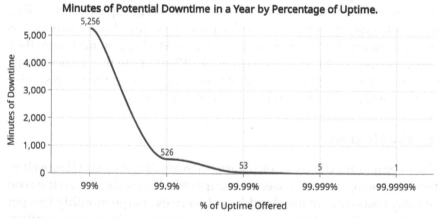

Figure 4.1: The potential downtimes compared by percentage of uptimes

It should be noted that part of that downtime will come in the form of scheduled maintenance and upgrades as the service providers further develop their product offerings. In the agreement with the providers, this is called *planned downtime*. The other kind of downtime, called *unplanned downtime* in contracts, would be from unexpected issues that the providers have on their end. These issues might include system errors, hackers, or force majeure. That's also where their scalability will have an impact on your company operations.

Performance

To achieve 99.999% uptime, service providers take advantage of cloud networking technologies. This allows them to share resources over multiple servers in order to keep their services operating smoothly. This comes in handy during peak surge times of usage such as a 9-to-5 workday on the East Coast of the United States. As more people come online to use a service, the cloud network shares the workload across multiple machines so that no single server will get overtaxed and go offline from the increase in usage. The backends of the companies are designed to scale to meet the expanded need as their user base grows. The bigger their cloud network is, the more users they can handle at any given time.

Part of the scalability factor is performance, and part of that is speed. An online service that is slow to load information is frustrating and wastes time. I've been on many calls with credit card companies or utility companies who apologize while pulling up my account, because their system is running slowly that day. If a small business said that often enough while managing customer questions or handling support issues, it could lose customers. You should be able to expect high uptimes as well as high performance from your service providers. Your business depends on both. Again, this is addressed at the provider's backend where they allocate resources to allow for scalability. Please note that issues which arise with third-party plug-ins that you integrate into the main service of the providers offerings will not count in the SLA.

Locally Hosted

Many companies choose to purchase software that they run themselves instead of using a cloud-based subscription. This method cuts down on monthly costs since many cloud-based services charge monthly fees per user. If it is software that you are hosting on a server in your own office, then you can get bottlenecked fairly easily during high use times. The server can only handle a finite number of queries at a single time. It's very much like the slowdown you see on your own computer if you have too many open browser tabs.

A software provider that hosts its application in the cloud can use a massive amount of resources to keep its service going quickly at any time of day or night. But that same software application, being hosted on your own computer or your server, is limited by the speed of those machines and the speed of your office network. It won't be easily scalable. If you do not have a large number of employees or if you have an IT person who is good at managing servers, then a locally hosted option may be a cost-effective solution for your organization.

A good example is QuickBooks. I know a number of small businesses that have purchased a version of QuickBooks Desktop Pro that is downloaded onto a computer that is dedicated to accounting purposes. A person can sit at that computer and work or others can remotely log in to that computer to use the accounting program from another offsite location. However, as these companies have grown, they've moved to QuickBooks Online. That is the SaaS cloud-hosted version of the accounting software, which comes with a suite of automated features as well as online support.

With the online version, you also get the assurance of not losing any important data. Understand that when you keep the software on a computer or server in your office, you run the risk of losing data if that machine is not backed up properly and something happens to it. If it is stolen or damaged in a fire or compromised by malware, critical information can become endangered and lost forever. With a cloud-hosted solution, that risk is virtually nonexistent.

Local or Cloud

Many people simply do not care for using a third-party software vendor. The reason being is that the practice can cause the company to end up in a position in which it is entirely dependent on the third-party vendor for its business operations. That is the heart of cloud-hosted SaaS solutions.

Some business owners would prefer to have their systems run independently of third-party-hosted systems. It gives them a stronger feeling of ownership in the software that they are paying for, and it helps them feel less dependent on an Internet connection for operations. They also feel that the software will be more reliable if it is downloaded directly to their local hard drive, similar to versions of QuickBooks Desktop Pro that we discussed earlier. But this does not help your company when disaster strikes and you cannot get into the office to use the computer that hosts the software.

SaaS history is long and well established. Figure 4.2 gives a brief history of SaaS services. Today these services and their ability to be accessed from anywhere is what drives remote work ability and will aid a company in disaster recovery and continuity planning.

Commercial Software Industry Timeline

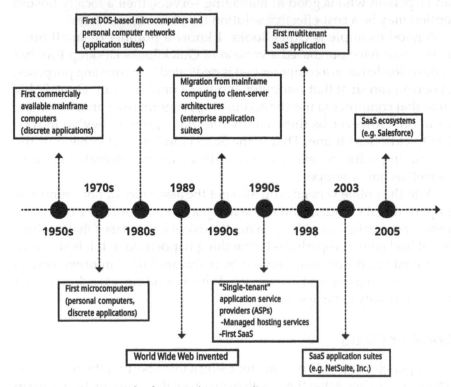

Figure 4.2: Commercial software industry timeline

Focus on the Need

When you are starting out looking for new tech for your company, you should focus on your need. What problems are you having that you are hoping this new set of services will solve? Let's again use CRM as an example. You might be shopping for a CRM solution to speed up your sales process, create stronger avenues of communication with clients, and have better control over your marketing. It may be great if the software can work with multiple areas of department focus, but you don't necessarily want to shop for an all-in-one solution. Sometimes a vendor tries to force-fit software and features to cover a lot of bases. When this is done, it's an attempt to beat out competitors who have similar suites of services. But you will find that with many companies, the add-ons aren't as good as the core product, so it ends up defeating the purpose of having a great overall product for their clients.

Since no provider can be everything to everyone, employing multiple software solutions is a better strategy. Each solution will have a strong primary function, with additional components that are not as robust as the main one. When I spoke with JP Holecka from POWERSHiFTER Digital, I asked him what services he used to take his team working remotely. Here is what he said:

> We use Google Workspace for docs, spreadsheets, presentations, and email. We switched (from) Invision and Sketch to Figma Cloud for design late last year, and it's been an amazing design and prototype tool. Slack is our go-to communication tool for internal and client-facing daily conversations. We now use Miro for all of our brainstorming and workshops, both internally and externally. Jira and Confluence Cloud for project management and intranet needs. Twillio is our interactive voice response (IVR) that directs calls to team members' mobile phones. Xero cloud for all accounting needs. Lastly, we use Harvest for cloud-based time tracking.

You may think that having so many different digital tools could be confusing for your team. The idea of having just a few core services is alluring as opposed to switching over to various multiple platforms throughout the day. This is a good reason to look at choosing providers that support integrations with third-party apps that you may already be using.

Integrations and Working Well with Other Tech

Look for services that work well with the suite of software that you already have in place or ones that you know you will be adding later. You will save hours of time daily for your company by choosing services that interoperate with each other. This resolves issues of having to manually enter the same information into multiple databases when new clients come onboard or as changes are made to existing accounts.

Smaller providers may only integrate with a few of the more popular business apps available, while larger providers will have an entire library of third-party apps that work well with their primary systems. Let's look at the leading CRM company, Salesforce. To help its customers with integrations of services different from their primary focus, Salesforce has the Salesforce AppExchange. Here, you can browse from hundreds of third-party apps based on the types of solutions needed, such as email marketing, payment processing, and even apps that are specific to industries such as nonprofits, real estate, and retail.

When looking at tech that works well with others, don't forget about the devices that your team will be using. You want to be sure that the software you use will also work well on mobile. Each year an increasing amount of business work ends up being done on mobile phones and tablets. That trend will continue to grow in the future as portable devices become more powerful and more convenient for business uses. You want to future proof yourself to optimize productivity by choosing mobile-friendly solutions.

Look at Your Future

I have done a good deal of consulting for commercial property developers who were looking at bids from various companies to build the telecom network for new facilities. In looking at the proposals from vendors bidding on the projects, I look for not only the current needs but the needs of the property in three to five years down the road as well. That is a good period of time for looking at your own plans for the growth of your company and predicting what it may need with regard to your tech. I prefer to cap the forecasting at five years because of the speed at which technology improves. It's impossible to know what kind of devices and new standards will exist after five years.

A good example is a meeting I had with a large national commercial property management and development group that I met with when I started my managed Wi-Fi company in 2002. Wi-Fi was very new at the time. It was not integrated into phones, there were few if any public Wi-Fi hotspots, and if you wanted to use Wi-Fi on your computer, you had to insert a wireless network interface card into the side of your laptop. I met with two senior VPs for the property development company and told them of the advantages that we could bring to their properties to improve tenant satisfaction. By adding Wi-Fi as a technological amenity and as a value-add to their conference rooms and common areas, they could ensure that the service would ultimately assist in their tenant retention as well as increased occupancy.

The two men smiled and said that several months ago they finished up a companywide project to install Cat5 Ethernet cables in the conference rooms of all of their properties. They included polycom conference phones and projectors in the upgrade. They assured me that their conference rooms were now state of the art as far as technologies go. They very politely added that although Wi-Fi was certainly intriguing, they doubted if it ever would be of much use for a meeting planner or business meetings in general.

You may not be able to predict what great new technologies or types of service providers might be coming in a few years, but you can forecast your company growth. Look at your growth plans and keep them in mind when looking at service providers and software. Try to see if these other companies will be able to be useful to you after some of your growth.

BEST PRACTICES

Look at the needs that you have today and look at where you plan to be in the future. Choose solutions that can be useful as you grow your organization.

Look for Reviews

When shopping for providers or software, look for reviews. You may find that companies post testimonials and case studies instead of reviews on their main website. If that is the case, do a search for reviews on their products and services. It's best to read unsolicited reviews from people who have used the product.

Case studies are great to help show a potential client what can be done with a suite of software services. But the main reason for the existence of case studies is for the company that publishes them to use them as a marketing and sales tools. From them, you will get real-world examples of how the software has been used to solve problems at other companies. But they will be worded by marketing-savvy people who will always show the software provider in the best light possible. They are still honest examples, but they are biased.

Thankfully, the industry has come up with a solution to finding relevant reviews to help you on your journey. SaaS listing and review websites are wildly popular among the tech community. These sites are specially built to show purchasers the top brands and providers in multiple industries. Many of these SaaS product review sites also help rank available choices as well as allow for side-by-side comparisons and, of course, user reviews. However, even with review sites, you need to choose well.

There are many sites that don't enact enough due diligence to ensure that the reviews are from trustworthy sources. As people can see from the popularity of Yelp and Tripadvisor, leaving and reading reviews have become a big business, so much so that some companies have employed bots and fake reviewers to leave kind words for their company and at times use them to bash the competitors.

So, if you use an SaaS review site, be sure to stay with some of the more trusted sites. On websites such as TrustRadius and Capterra, the reviews are vetted, and they verify that the reviews are coming from real people who have experience with the software as opposed to people who are paid to pose as customers. They do their best to eliminate bias and marketing ploys from their sites. This is all designed with intention to help you build more confidence in your choices for a new SaaS provider.

Compliance

If your company specializes in certain fields such as medical, legal, or finance, you may be required to use SaaS vendors that are certified in specific compliances. For example, if your company is in the medical field, you may need to find a cloud data storage company that is also compliant with the Health Insurance Portability and Accountability Act (HIPAA). Simply having that HIPAA compliance emblem on its company website is not enough to ensure that the services you choose from that company are actually compliant. It is up to you to do due diligence and ask if the specific suite of services that you need will be compliant. And there will be further steps that you need to take.

You will need to get a signed business associate agreement (BAA) between your company and the cloud storage company where protected health information (PHI) may be stored. This is an agreement that creates a bond of liability between the two companies as healthcare data is being passed between them. This also will help satisfy HIPAA regulations for your company and help to keep you compliant. Be aware that simply using HIPAA-compliant vendors does not make your own company HIPAA compliant. It is up to you to make sure that your organization satisfies all needed regulations to ensure compliancy.

First-Time Setup

One of the factors that you will need to weigh when looking at software services is what it will take to get you started. By that, I don't mean just the cost of the service (which we will discuss in detail later in this chapter); I mean that you need to be aware of steps you will need to take from the point of purchase until your team is using the new system as intended.

In order to be able to plan accordingly, you'll need answers to a number of questions: Is this platform intuitive to the point where your employees will be proficient users in a just a few days? Or is this something that they will need training to use and it will take a couple of weeks before they are accustomed to their new work tool? How long will it take to have the new system configured in the way that you are getting the most out if it for your company?

Configuration

When you start your initial setup, you are going to need to have a good understating of the steps involved. In some SaaS solutions, the entire platform is hosted in the cloud. With that type of service, you may not need to download anything onto your desktops. When everything is hosted in the cloud you won't lose any drive space on office computers. In that case, once you set up your hosted company dashboard, everyone in the company will get the same user experience. So how long will that take? That is going to depend on what the service is and how it will be used. Again, we will stick with a CRM solution to use as an example. Before you begin to import customer data, you will need to configure and set up the data fields in the CRM software to match the info that your company uses and will add into the new system.

You will also need to look at how some departments in your company may use the service compared to other departments. This will allow for multiple configurations based on the department. Perhaps support will use some tabs and information that marketing will not and vice versa. Configuring by department will empower those individual teams to work more efficiently and in a less cluttered dashboard. This user-based configuration will allow for more customized reporting and streamlined page layouts that create a better workflow for the designated tasks of each team.

This type of initial configuration will cost you more time to set up and get your company started. However, it will be worth the investment of time in the long run as your team works more efficiently and cohesively on the goal of growing your company.

Customization

Although you customize your configuration to suit your company, customization means something different than configuration. Customization is altering the service's code to meet your business needs. This could be the use of Cascading Style Sheets (CSS) to alter the page layouts and appearance. Another example is adding third-party integrations that will help you use the system in ways that you cannot without those add-ons.

This is when you will want to check and test your integrations for compatibility and interoperability. *Compatibility* is when different applications perform in the same environment. For example, let's say you have your email integrated into your CRM system so that you now send and receive email via the CRM system. You want the email to function as well via the CRM system as it would via Gmail or Outlook.

Interoperability is how a system functions when interacting with other systems. Let's say that you have your CRM system set so that when a sales team member enters a new customer into the CRM system, it automatically sends that customer data to QuickBooks where the accounting team will be able to use it for future invoices. The full and accurate data transfer into QuickBooks would be a test of the systems' interoperability with each other. In most cases, you will want to test that the data will flow well bidirectionally between apps. This way, when new information is added to an account in one system, it will show up in both systems to save time from team members inputting the same data more than once in multiple systems.

Administration Levels

You will want to look to see what administration and accessibility levels can be set for managers and users. You don't want a normal user to be able to change field designations or delete sections. That could cause havoc and unintentionally erase hours of work. You want managers to be able to add on new hires and allow them to provision login credentials for the new workers. You also want managers to be able to disable access immediately when an employee leaves the company. But you may want to leave the ability to make customizations strictly as admin-level privileges since those customizations will likely make changes globally across the system.

Assigning those roles and privileges to your team will be part of your setup and getting started. I consider accessibility levels as part of your security as well. You want to be familiar with these and other security features. Ultimately, it should be something you ask prior to committing to a vendor. For example, are communications through its services encrypted? Does the vendor offer multifactor authentication?

Data Migration

Before you can move data from your other systems into the new one, you will have prep work to do. Prepping is arguably the most critical part of getting you started using the new software. The success of migrating the data is dependent on your prep work. Don't rush it. Just as moving from one house to another takes planning and work, moving data properly takes planning and work.

Ask your vendor for a realistic estimate of how many hours your team will need to dedicate to this task so that your migration goes smoothly. Be sure to set aside enough time and team members for this part of the task.

Data Selection

In this phase, you will look at what data you have in the current systems and which data sets you will move over to the new system. This is your opportunity to get everything into the new system looking exactly the way you want it to look. Consolidate and standardize client and company data fields. This keeps information easy to read at a glance.

Also, when prepping the new system, create relevant fields and limit the number of free-form fields where team members might be able to put in any information they'd like. This stops people from adding important info into notes sections that should be added to standard fields instead. This will again save your company time in the future so that others won't have to dig to get relevant information.

Get input from your team when you are deciding what data fields are important. For example, see what data fields sales and support might need, then merge any fields that can be put together and eliminate any that are not needed. This will also save data entry time for future contacts.

Data Formatting

This is a good time to fix formatting mistakes so that they don't drag over to the new system. You want the new system to be clean and fresh, not bogged down by old mistakes. A good example is capitalization of names. It may not sound like a big deal if this is a system that will be used internally, but this type of data tends to be used in other systems such as marketing. Sending out personalized emails with the first or last name misspelled or in all lowercase can be a turnoff to the recipients.

Check for consistency in your existing data. For example, phone numbers can be entered in different ways by different people. Do some use hyphens between sets of numbers and some use a string of 10 numbers? Do some use parentheses around the area code whereas some do not? Another data field that is often inconsistent is job titles.

Inconsistent data entry can have a detrimental impact on your return on investment (ROI) of the new system. For example, *chief executive officer* and *CEO* mean the same thing. However, if you were to run a targeted email campaign and only enter CEO as a target parameter, anyone labeled as chief executive officer may not get the communications from that campaign.

BEST PRACTICES

Data integrity is crucial for effective sales and business operations. Train your team to be consistent and to strive for accuracy when inputting information into company systems.

Data Deduplication

Deduplication means getting rid of duplicate data. Taking the time to do this will save you money. For starters, it lowers your data storage costs. The more data you have, the more storage costs go up. A system bloated by duplicate data costs a company more in storage fees, and it costs more in time to migrate the data, do backups, and run reports.

Also consider the time wasted if a client is duplicated in the system and people from separate departments are each continuing to enter information on the same client but in two different profiles. It complicates both sales and support if no one can see the whole client history.

Many systems have features that allow you to run reports and look for duplicate info, merge the profiles, and delete the redundant ones. All of these steps will help with your overall data integrity. Data integrity is the accuracy and consistency of the information that you have in these systems. They are crucial for efficient business operations. You should emphasize this to your team so that they make data integrity a priority each time they add information to your company systems, not just during a migration process. Standardize how new information is input across all departments.

Mapping and Templates

The mapping process will help you decide what information is going to be moved to the new system and what can be left out. During this process, you will tag fields so that they are identified in the current systems and information is placed in the correct corresponding fields in the new system. For example, you will tag and map fields such as first name, last name, office number, direct extension, and so on. Therefore, you will want to decide which fields will be *required* and which will be *optional*, and what fields the new system itself will require you to have.

Once you choose these fields, you will be able to create a template that will speed the process up considerably. Oftentimes, this will be done in Excel. However, many SaaS companies have their own mapping features to help you prep and tag your files to transfer over to the new system. After you have the template created, you will review the data for each field and then populate the fields. Start off with loading a single record or customer file to the fields in the new template. Make sure the data matches the fields that you want and check to see if all relevant info was transferred to the template. If your results are good on the first record, load the rest into your template and then you should be ready for migration.

Migrating

How long it takes to migrate everything over will be determined by the volume and complexity of the data. This is another good reason to cleanse your data and get rid of duplicates and out-of-date records. If things go wrong with the migration, don't worry. You can usually run the migration several times if need be. This helps you fix migration issues that you may find with fields or validation issues.

Outside Implementation Companies

I'm sure that to many people, all of this sounds like a lot of work and a lot of time. It certainly can be intimidating, and that is what causes people to hesitate pulling the trigger on the purchase of a new CRM or other type of business service. This is why third-party implementation and integration partners have come in handy for many small businesses.

These integration partners specialize in consulting on and integrating specific SaaS products into organizations. By doing so, they empower the small businesses to use the new service optimally, and they speed things up from time of purchase to getting teams fully functional with the new product. This will help the business to make the most use of the new services and make sure that it is set up and installed in a way for the company to best leverage its features and increase its ROI. If you think you may benefit from such third-party integrators, then you should include that in the calculation of the cost and budget for your new tech services.

After the Sale

When choosing a provider for your company, you should look at how you are treated before the sale because it's a good indicator of what your experiences may be like after the sale. Note whether they were responsive to your questions and answered them fully or gave somewhat vague answers. You don't want to be left on your own should you need help. Here are some things to consider to help you make the most informed decisions on a new software service.

Customer Service

There is a big difference between customer service and customer support. Customer support teams tend to focus on resolving technical issues. Customer service teams are there to help build a good relationship with you after the sale. They are there to advocate for their brand and help retain customers. They do this by making sure that your needs are met and that you are able to happily continue using their company. You should think about this when weighing vendors and service providers for your company's tech needs. Ask what happens after you have signed up with them and what kind of customer services they offer post sale.

A product or service with strong customer service will save a business both time and money. A good way to understand this is by looking at companies that offer little in the way of customer service. These are companies that direct their client base to have more of a self-serve experience. They typically have all of their documentation online, so should you have a question or concern, they will direct you to these documents. You will also see many of them have chat windows on their websites where bots will do most of the communicating with you. These bots will give canned answers and also direct you to their library of online documentation or even to the forums that they host for other end users and customers of their products and services. In that case, it is the community of end users that will be more helpful in answering questions, based on their own experiences. That is the basis of community-driven support.

Handling customer service in this manner benefits the company more than the customer. It takes the pressure off the company to answer questions and from interacting with you, as the customer. It saves them money, but it does not make your experiences working with them very easy. In these cases, you will be the one who has to research the answers to your questions and concerns. In other cases, a provider offering good customer service will have the same self-help resources available *and* have people who help you during the whole of your customer journey.

Customer service reps will not only help you get started but also help your company grow with their brand. They do this because their position is dedicated to creating potentially lifelong customers. They can assist you when it comes time to onboard new people in your company, and they will reach out to you to let you know when new features or integrations are available that would be beneficial for your company. This kind of real customer engagement leads to higher customer satisfaction and stronger retention. You will appreciate the difference good customer service will make if you have ever experienced *bad* customer service from a business-to-business company.

Customer Support

When you have technical issues, you will want to reach out to the vendor's support team. Today, technical support is handled by communicating in a variety of methods. Before you buy, you will want to know what to do should you have technical issues and how you contact the right people, the ones who can help resolve specific issues as quickly as possible to keep your work flowing.

Many companies offer a help desk for their customers for basic support needs. These are usually staffed by tier-1 support personnel who can handle most of the simpler technical issues. These are issues such as resetting a password, helping to navigate the company user dashboard or website, or answering how-to questions. In most cases, these help desks are not staffed by overly technical people. Many times, they are using a script and online manuals to help you resolve your issues. If you ask question A, they respond with the corresponding scripted answer and walk you through the steps of resolving your issue by following the script they are given. They are good for the most frequently asked support questions, but they are very limited beyond those issues.

These support teams may not be available 24/7 and that is something that you should be aware of ahead of time. There may be a toll-free number that you can call, or more likely they will prefer that you use a widget on the company website to help you with your support issues. These can be a combination of bot and human support agents. The bots will direct you to self-help documentation or be pre-programmed to give canned technical answers based on the key words of your questions. If they are not able to help you, they will direct a human to join the chat, but only if one is available because they likely work during set business hours. Some vendors keep those support teams working limited hours for financial reasons, so you may not reach a live person until the next business day, depending on when you reach out with your issue. Keep that in mind if you or your team do a lot of late-night work. Working late is not uncommon for developers or creatives who tend to prefer late-night hours.

If your technical issue is beyond the tier-1 team, that's where tier-2 support comes in. Tier-2 support technicians offer more in-depth support. They are very knowledgeable about the systems they support, so they are technically able to resolve most issues that you and your team may encounter. They have access to deeper subsystems and a larger knowledge base about the products and services of their company. Usually, you would not get to a tier-2 support technician until after you have passed a tier-1 support person. Also, tier-2 support people are most likely not available 24/7, although some larger companies keep at least one tier-2 support person available at all times for business-to-business services. That's part of the difference between consumer-level services and business-level services. Business-level services will have a better support structure for their clients.

Response Time

When looking at providers, ask about their average response times. A response time is the time between when you initially report an issue and the moment that a support agent performs some sort of action to help resolve the issue. It may be hours before your problem is actually resolved, but that's a measurement of the resolution time for your problem. When you report it, you want some sort of response that acknowledges that your problem is being worked on.

Many support issues are still handled via phone calls. Your time is important, and you don't want to spend a good deal of that time on hold waiting for help. How long it takes to resolve your issue will depend on the issue itself. However, the longer it takes to get a response from support, the longer it will take to get someone working on your issue. So response times are an important factor in your overall satisfaction as a customer. If you and your team will end up being unable to use the system for several hours until you have a resolution, your company will lose time and money. Faster response times means that your company is up and working more quickly.

Total Cost of Ownership

The total cost of ownership (TCO) of new company software is something that many small businesses fail to fully calculate. It's rarely just the monthly per-person licenses that cost your company money to use these tech services. To get a more accurate ROI, it's important to gauge the expenses that you will put out over the lifetime of use. Understanding the TCO should be one of the major factors in your decision-making process when choosing services and software vendors.

Even software that offers free basic versions can have a financial impact on your company. Your team may use some freeware for several weeks or months and find that your company growth has changed your needs. That free software did not come with any direct costs to you, but there were indirect costs. Your team slowed their work down to learn to use that software and integrate it into their daily operations. They spent hours inputting data into a system that did not ultimately meet your company needs. Now they will start the process over again with a new vendor, new processes, and new software. In the long term, freeware will cost you time and money.

TCO for SaaS Applications

Summary of the cost allocations of a SaaS deployment

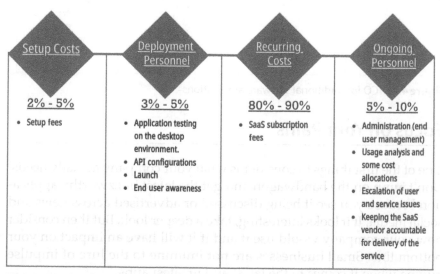

Setup Costs	Deployment Personnel	Recurring Costs	Ongoing Personnel
2% - 5%	3% - 5%	80% - 90%	5% - 10%
• Setup fees	• Application testing on the desktop environment. • API configurations • Launch • End user awareness	• SaaS subscription fees	• Administration (end user management) • Usage analysis and some cost allocations • Escalation of user and service issues • Keeping the SaaS vendor accountable for delivery of the service

Figure 4.3: TCO for SaaS applications

TCO for Traditional Software Applications

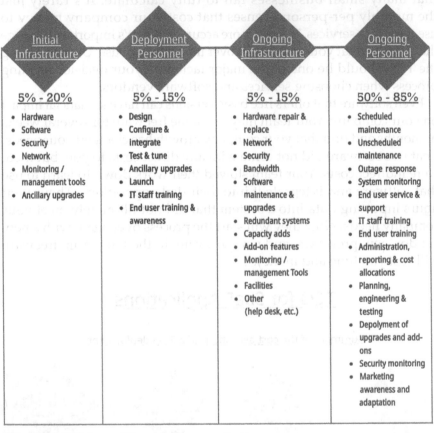

Summary of the cost allocations of a traditional software deployment

Initial Infrastructure	Deployment Personnel	Ongoing Infrastructure	Ongoing Personnel
5% - 20%	**5% - 15%**	**5% - 15%**	**50% - 85%**
• Hardware • Software • Security • Network • Monitoring / management tools • Ancillary upgrades	• Design • Configure & integrate • Test & tune • Ancillary upgrades • Launch • IT staff training • End user training & awareness	• Hardware repair & replace • Network • Security • Bandwidth • Software maintenance & upgrades • Redundant systems • Capacity adds • Add-on features • Monitoring / management Tools • Facilities • Other (help desk, etc.)	• Scheduled maintenance • Unscheduled maintenance • Outage response • System monitoring • End user service & support • IT staff training • End user training • Administration, reporting & cost allocations • Planning, engineering & testing • Depoylment of upgrades and add-ons • Security monitoring • Marketing awareness and adaptation

Figure 4.4: TCO for traditional software applications

Focus on Your Pains

One of the first things to consider is what your company actually needs. Don't jump on the bandwagon and get the latest buzzworthy application because you see it being discussed or advertised across news and social media. If it looks interesting, take a deeper look. But then consider how your company would use it and if it will have an impact on your bottom line. Small businesses are not immune to the lure of impulse buying when it comes to the latest and greatest apps.

If you are shopping for software, ask yourself what problem it is intended to solve for your team and for the company as a whole. Figure out what pain it is going to relieve for your organization. If you can't answer that question, then you may need to ask if the software is worth the time and manpower to learn and implement. Of course, you may be looking at your growth and looking at software that will enhance your ability to handle a larger amount of business in the future. If that is the case, then look at all of your current processes. You will want something that will positively impact your ability to grow, but if these other processes will not be able to grow alongside new processes and software, you may be in for a larger purchase altogether.

Future-proof your company by looking at your growth plan and making sure that any new software you implement will be able to grow alongside of you. Growth means a larger team and a larger customer base. More team members usually means more software licenses. Some software services also limit the number of customer data entries for certain pricing levels. That means that your total cost of ownership will increase as your company grows. That is also a relevant factor when calculating both your TCO and your ROI.

Upfront Costs

The upfront costs to get your company going on new software is the front end of the set of costs to take into consideration when you begin to calculate your total cost of ownership for your new software implementation.

Software

Upfront costs are going to vary by a large degree if your needs are not something that can be met with an out-of-the-box product—one you can buy from a retailer or SaaS company that has all of the features and services your team will need. Whether you can use an out-of-the-box solution is not something you will discover when you first look at the real needs of your organization. Once you know that, then you can shop intelligently.

You may find a number of solutions that come very close to meeting your needs but don't quite fit. This is where integrations come in handy. In this situation, you should narrow down your vendor choices to those that allow for integrations with third-party vendors. If this is the case, then you should be able to find something that very closely meets your needs with some plug-ins and integrations. Look for software that works well with the software you already have in use. You will save hours of time and money with data entry alone.

In some cases, you may want a semi-custom build. This where you will find software that you really like and has most of the features you need but is still lacking. There are software development companies that are very knowledgeable with certain industry leading software and can create customized add-ons for your company to make the main software be all that you want it to be.

Software companies realize that they cannot be everything to everyone, so they make their applications programming interface (API) available to other developers. The API acts as a gateway to the back end of the main software application so that developers can go through the gateway with their own application and have them work together to achieve a task that neither could do on their own. You need only ask the software company if it has any development partners it already works with to create something for you and get you started.

Hardware

If you are using an SaaS product that is cloud based, you may not have any real hardware needs. Everything will be hosted for you online, so you don't need to be concerned with your network. Does your team have the bandwidth and Internet speeds they need to access online systems and not experience much lag, or will they become frustrated and end up not wanting to use the system due to periods of slowdown? When they are in the office, do you have a fast network with Gigabit Ethernet switches and up-to-date computers that have plenty of memory to avoid bottlenecks? One of the biggest reasons people stop using some office software systems is because they are slow and cause workers stress. This means that the software investment does not get the usage it should because it was not positioned as it should have been in the first place.

Make sure your network and hardware can handle what they will need to before you invest in the new system. This is especially true if you are hosting the software on a server yourself. Regardless, if that server is in your own network or at an outside location, your internal network still needs to be able to handle the traffic back and forth. During peak hours, a slowdown will cost you time and money and perhaps some clients if it affects the quality of work.

Implementation, Migration, and Customization

Depending on your needs and on the size of your company, how you get started might involve a cost that you have to put into a budget. In many cases, getting started with new software could be as simple as your team logging in online and getting to work using a cloud-based dashboard. Other times, you will want to add browser extensions, mobile apps, and desktop downloads. And there may be times when you want to involve a third-party team to get you started.

You will probably want to invest in a third-party company for implementation if you need a large amount of customization. Many times, even if the system is user-friendly for nontechnical people, it will cost you in time to learn the system and customize it to be more user-friendly for your company needs. Implementation companies can not only save you time, they can guide you in adding some features and integrations that you may not have thought of using.

These companies will help you integrate any systems that you may want to interact with the new software. Many times, they can also help in developing any custom integrations that you may need now or down the road. Keep in mind that you will want to keep a good relationship with companies that do any custom development for you. There will be occasions when the main software is updated and the custom add-ons need to be tweaked in order to work with the new updated version. It behooves you to find a company that is responsive, will be easy to work with and will still be around in a few years should you need support with your custom code.

Hiring a third-party company to migrate data from older software will also be worth the cost if you have a large amount of data that needs to be moved over. It will help you make sure that all relevant usable data is brought over to the new system as well as help you clean up that data and make sure it is plugged in to the new system in the right places and in the correct manner.

User Licenses

If you are purchasing something that is cloud based or out-of-the-box, you are most likely going to have to purchase additional user licenses for the team members who will also use the service. These license purchases can add up quickly. If they don't offer, ask if there are any discounts available if you pay annually versus monthly. Also, ask if they offer any discounts for your industry, especially if you're in education or work for a nonprofit. You'd be surprised at how many companies are willing to offer even a slight discount if you simply ask for it.

BEST PRACTICES

Have a designated project manager in charge of new software purchases. Ensure they look at all costs involved for startup and continuing operations.

Training

With some systems, the design is so intuitive that there is no in-depth training needed. There will still be a slight learning curve, and there will be a short period during which you will have to work it into your processes to ensure that it becomes a habit to use the new system. Expect the transition to slow down your team's productivity slightly for a couple of weeks while they get used to the new processes.

Other times, you will want training. There are multiple options that will be made available to your team. Many times, the free self-help and self-training will be enough. They will be complete with documentation and online videos to get you started. In other cases where the system is very robust, they will give you a set number of hours of onboarding and training that come with your new purchase. In these cases, they are always happy to sell you additional training hours if you feel the need.

Operating Costs

Your operating costs for the new software are the next items that you should take into consideration when calculating your total cost of ownership. These are the day-to-day expenses that you can expect when your teams are using the new system.

Software Support and Maintenance

Software needs constant support in the form of patches and updates. In many cases, these will happen automatically during off-peak use hours and without much loss of performance for your team. However, with very robust systems, you may need added support, especially if there are any custom third-party integrations.

You can purchase annual support agreements from the vendors with a set number of hours included with each level of support package. Be sure to ask if upgrades are included with your software.

Additional Licenses and Training

As you grow your company, the new team members will also need access to the relevant systems. Keep in mind the cost of system training and user licenses when you consider onboarding new employees.

End User and Admin Tech Support

If you don't have an IT team in your organization, you may want to have third-party support available. Troubleshooting may take a more technically savvy person. Many cloud-based out-of-the-box SaaS products will offer self-help and limited support at no charge. For more in-depth support, you will need to pay out of pocket.

Network and Data Center

If this is software that you will be hosting in-house or at a data center, then keep those costs in mind. Maintenance of the hardware and network as well as monthly broadband will be part of your ongoing costs. Also be sure to add in updating those systems when security patches and other updates are available.

Backup and Disaster Recovery

For most SaaS systems, you will not need to worry about backing up your data. However, anything that is on your computers or on your server will need to be backed up on a very regular schedule. That data is precious, so don't skimp on protecting it.

Scaling Up or Retirement

Hopefully, your company will be growing for long enough that you will need to look at scaling up or retiring software at some point. If your company experiences so much growth that this new system no longer fully meets your needs, what would it cost to scale up to an enterprise version, if that is what you will need? If it does not scale up any higher, you need to find out what it will it cost for you to retire this service and migrate to something more robust. There will be times when you have to run the old system side by side with the new system for months while incurring costs for both.

Choosing Wisely

As you see, there are a good number of considerations for a company to look at when getting new software. You will choose well if you focus on your needs first along with where you see your company going down the road. Plan for growth and the needs that may come with it. You want to choose software partners strategically. Get ones that are reliable with excellent reviews from real customers of their products.

Take a good look at all of the costs that may be involved. You do not want to just settle for something that may not meet all your needs but is close to meeting them. There are many companies that can build you what you want from the ground up. In the following interview, I speak with CEO Marc Aptakin of MAD Studios. MAD Studios develops highly innovative software solutions for its customers, which can be a great advantage for companies with more specialized needs. There is no need to stick to just out-of-the-box products. Finding or creating the right solution will save your company a lot of money and time. Managing your time well is always a smart method to increase a company's bottom line, and we go into that topic in the next chapter.

MARC APTAKIN, FOUNDER AND CEO MAD STUDIOS

Company Profile

- Location: Miami, Florida
- Employees: 110
- Primary Line of Business: Manufacturing solutions and a variety of marketing services and software solutions; creative reality services like virtual reality, alternate reality, and mixed reality.
- Primary Audience: MAD is a business-to-business organization. We offer two different components: agency and manufacturing. Through our agency, we provide creative solutions for businesses through the use of services such as marketing campaigns, social media campaigns,

and rebranding initiatives. On the manufacturing side, we offer businesses custom packaging, fulfillment, displays, and other similar services.

Brief Company History

MAD was founded in February 2001. It was an offshoot from a partnership split—an amicable one—from another agency called Vantage Design. We founded Vantage in 1996 and split up in December 2000. From there, MAD was born.

What apps, services, or technology did you use to bridge the gap from in-person to remote in order to keep the workflow alive with your team?

We used a variety of platforms to stay connected with our team. We frequently use Basecamp, Google Hangouts, Zoom, Slack, and Microsoft Teams. With regard to the client, we try to stay accommodating to whatever the client prefers using.

What was the most difficult part of going remote for you and your team? Did poor Internet connections or slow computers owned by the team affect you?

The collaboration was definitely affected the most. Being in the same room and joking around oftentimes leads to great ideas. When things turned remote, I could feel a lag in the synergy. Poor Internet connections and slow computers did not help our case either. We noticed people getting frustrated and not being as engaged, which was a big obstacle during COVID-19.

What was the easiest part of going remote for you and your team?

As a team, everyone here was rather technologically savvy to begin with. A lot of these kids wanted to work from home anyway, and this gave them the opportunity to do so. On top of that, the oversight from the management did a great job of staying on top of deadlines. Making this work was a huge team effort, and we really came together and made it happen.

How did you work remotely with clients?

The agency side of the business is pretty standard. With features like Zoom screen sharing and iMessage, sharing information with clients virtually is easy. You're just dealing with a little bit of extra back and forth.

With regard to manufacturing, prior to the pandemic, the client would have to physically come to the warehouse for an audit. But during COVID-19, we did the audits virtually by walking the warehouse with a client on an iPad. This method worked a lot easier than we thought. It saved them a lot of time, and many even said they'd prefer this method in the future.

How did working remotely affect your team working together? Did you notice much difference in how your team worked together? Was efficiency affected? Was the communication of ideas affected when done remotely?

Communication and idea generation were both affected. About a year before the global pandemic, we brought in Roy Hudsell and Rafa Ribeiro, who helped the low and middle levels of the agency "up their game" and increase their thought processes. When MAD suddenly switched to remote work, you could see people who were on an upward trajectory slide back a bit and get complacent. Now that MAD is allowing people to trickle back into the office, productivity is visibly different.

What would you do differently if you had to again suddenly switch your organization to working remotely?

If MAD went completely virtual, I'd add some of our own technology that's specific to how we work to try to bring some of that collaboration back. We believe this place is different from others, yet we are using the same tools that other agencies are using. We would need to build a custom program to grow that teamwork and synergy on a virtual level. We would implement wellness programs, team happy hours, and other events that would help everyone feel connected and united. Whether we're solely virtual or operating in house, connectivity and collaboration are the keys to success.

What words of wisdom and advice could you give to someone else who finds themselves having to suddenly make their operations go remote?

My biggest piece of advice for the management of an organization is to make sure your team members are staying on track and meeting deadlines. Be a resource! With the proper structure in place, your business can continue to be successful.

Time Management: The Result Is What Matters Most

Time management during the crisis that causes your company to switch to remote working will be a challenge. It's a distracting period, and during work hours you will have your focus divided between your normal day-to-day responsibilities and crisis management. This will also be a difficult period with regard to time management for your team and co-workers as they muddle to find a balance in their situation of suddenly working remotely.

Proactive steps that we discussed in previous chapters will be a huge advantage in managing your time well. In this chapter, we will touch on best practices for you and your team to keep better control of your time as well as some tools you can utilize to gain some advantages toward reaching your goals.

Being Productive

People discuss being productive from different perspectives. You may often hear someone say, "I've had a really productive day." By this, people usually mean that they have completed a number of tasks during the day. It can also mean to some that they have done a number of things in general. To someone who has been bedridden for an extended period due to sickness, just getting up, showering, and getting dressed on their own can be seen as a highly productive day. At the same time, to a highly active and healthy person, getting up, showering, and getting dressed would not even register on their scale of productivity. What I'm getting at is that *productivity* is a subjective term and there can be a disconnect in what being productive means in various situations.

When your company is flung into the situation where everyone is suddenly working remotely, it's not going to be a typical workweek, month, or quarter. The whole experience is going to skew your idea of what the current quarter and next quarter will look like for your company. You will need to develop new ways to measure productivity. You will need to train yourself to look at how people work in a different light so that you can guide them and yourself to overcoming the remote work obstacles.

Measuring Productivity

If you were managing a conveyer belt of robotic arms and other machinery, your measurement of productivity would be simple: output ÷ input = productivity. It's a formula to calculate your production processes. The input would be raw materials, the work, and the time to complete; the output would be the finished product. If you are not running an assembly line, you may not have such a direct method of measuring productivity when it comes to your business.

As we have discussed in previous chapters, simply being busy is not the same as being productive. Therefore, measurement by direct labor is oftentimes not an appropriate method of evaluation. In remote work situations, productive procrastination becomes a work killer. Working alone and away from your team can allow for minor tasks to become huge distractions that move your focus away from what you should be doing. If you did not have a solid established metric for measuring productivity in the office, it will be difficult to create a fair method once you hit a crisis that forces you to go remote.

There will clearly be changes in productivity. This is part of the adjustment to each individual's work environment and being separated from the team. If you were dealing with robotic arms only, putting them in a new assembly line would produce roughly the same output as in the old one. When dealing with people, however, you need to remember the human factor and be prepared to give some leeway as everyone acclimates. This includes yourself.

I spoke with Wendy O'Donovan Phillips, CEO of Denver-based company Big Buzz, about her experiences in leading her company during the shutdowns of COVID-19. I asked her what aspects of her personal management style needed to be altered in the sudden shift to suddenly working remotely. This is what she shared in response:

> I am a driving, focused manager: do the task, solve the problem, finish the plan. In this day and age, I now better understand the softer side of leadership. It's important that I listen to my team, be empathetic to the challenges they are facing during this difficult time, say and do the right things to unite them and build a culture of trust. Those soft skills are actually more important than the hard skills. When people trust the organization, leadership, and each other, they produce better work.

When your crisis first hits, not seeing your team in person will cause some anxiety and trepidation about the work being produced. This will happen in most cases even if you are not one of the types who need to see people in the office in order to manage them. It's the anxiety of the crisis that will drive the feeling that you need to ensure that all of your bases are being covered. Many interviewees switched into a "nose to the grindstone" mentality. They saw the crisis as a threat to the business and its productivity. It's a natural reaction to an emergency. Many will want to eliminate any immediate threats and guard against what else may come that could cause further instability. This mindset could easily endanger the trust that you have with your team if you are not careful. Trust and your company culture will best guide you during the times when your anxiety levels are running high from crisis management. When your anxiety related to crisis management rises, think about what you can control and shift your focus there. That will help you quantify the productivity of the team's remote workflow. You may want to use some of the simple techniques covered in the following sections that give you an added ability to maintain focus.

Objectives Focused

In Chapter 2, "The Remote Workplace: Set up Your Mind and Your Space," we discussed the importance of goal setting. It helps create short- and long-term destinations to steer your work toward. To keep up productivity, make sure the team, including yourself, is aiming at specific targets each week to arrive at the place you need to be in order to accomplish your goals.

With everyone essentially working alone, they still need to be accountable to the rest of the team. The inability to be physically present with each other means that team leaders must set daily and weekly targets that help accomplish the larger goals. It is up to the leadership to see the whole picture, be aware of the objective, and keep all team members on the right paths to get there. This is where your daily check-ins can come in handy.

Check-ins and Target Adjusting

When you have your daily or weekly check-ins, leverage them to best keep up your productivity. Ask about the targets that are set and see where everyone is on their deadlines. The leader will keep track of everyone's progress with the larger objective in mind. What is very important is to get feedback. Is everyone getting what they need from other team members in order to facilitate them to reach their targets?

BEST PRACTICES

Be aware that the crisis and managing it will cause a good deal of stress on leadership. Do not allow it to endanger the bonds of trust in your team.

Hold people accountable to themselves and each other. If someone is not producing, reset the target if need be, but investigate further to see where the bottleneck is occurring. It is important for leadership to be a resource and a problem solver as opposed to being a finger pointer. See what more *you* can do to keep things on track. You may need to reallocate resources or redistribute some of the workload.

Again, keep in mind that the team is not working in their normal environment. Occurrences in their home, where they are now also working, may have influence on their ability to be productive. That's part of the human equation when it comes to remote working productivity. Do private, one-on-one check-ins with your team. Be empathetic and be available to listen and understand what may be going on with them and in their lives. Then, if need be, adjust the targets to empower your people to be able to meet the bigger objectives.

How you do your check-ins is a matter of personal choice and management style. The method—phone, video, chat—is less important than the time you invest. Some companies have a dedicated Slack channel just for their check-ins.

Employee evaluations are also important because they can identify the employee's strengths and where they have room for improvement. Think of aspects of the check-ins as micro evaluations. The team needs to hear where they need to refocus some energy and where they are excelling. Validating an employee's good work helps them take pride in the work they are producing and helps keep morale up. It can also establish transparency among employees' work, which can strengthen the foundation of trust.

Deadlines and Prioritizing

Deadlines, even artificial ones, are a critical tool in maintaining productivity and reaching goals when the team is distributed. The deadlines help maintain accountability and can ensure that employees reach the goals set for themselves and for the team as a whole. Use them strategically as a method to keep up productivity and manage time use.

Team Deadlines

Involve others in setting the deadlines for the remote teams on projects. Unless you are the only person who interfaces with the customer, setting attainable deadlines should include input from the group. Employees will each have different sets of knowledge and perspectives on what is needed to reach goals. Set the team up to succeed by getting their feedback and input in establishing and adjusting deadlines.

Deadlines for Yourself

It's easy to get distracted by the crisis and your new work environment. Something that I prefer to do is to create artificial deadlines for myself. When I have a firm deadline for a whole project, I often create smaller, flexible mini-deadlines to get the various pieces done. Once I put those pieces together, the rest of the project takes shape.

Time Boxing

To help me manage my time, I box out periods of the day for different tasks that have to get done that day. I make appointments with myself to work on a section of a project. When I have huge projects in front of me that may cause me to procrastinate, I will change my viewpoint to make it appear more manageable. I often envision the project as broken up into small projects. Each smaller project is far more manageable as far as my time and personal resources are concerned. I prioritize each one in the way that they impact each other so that one task leads to the next, almost like dominoes.

This allows me to stack them in order so that I can set them off for the finale when I have each block in place. I box out the time for each one, and if I don't complete one in the time I give myself, I still move on to the next piece when that next box of time comes up. I approach it like a standardized test: If I get stuck on one problem, I move on to the others and come back to the problem question at the end. This allows me to complete far more in a single day than if I allocate more time from another box to the piece that I am stuck on. As the saying goes, you can't eat an elephant all at once.

Productivity Apps

Productivity apps can help you manage your time, but there are many different types that focus on various elements that influence your day. I will discuss them in separate disciplines according to the strength of the particular software. Regardless of whether you are a solopreneur who runs your business on your own or the head of your company leading numerous teams, you may need a little help maintaining an edge when it comes to staying productive.

Only you can decide what parameters determine your level of productivity during your crisis. However, efficiency is a less subjective measurement. You can take definitive steps that have an observable and a measurable effect on efficiency—in other words, on how you use your time. The sudden switch to having everyone work remotely is a new experience, and the same old efficiency tactics may not work as well in a new situation. You may need an extra push to maintain efficiency and productivity. Thankfully, there are some apps for that exact need.

Calendar Apps

One of the best indicators of efficiency is how you spend your time. Spending time in a productive manner can be as simple as proper scheduling and managing your calendar. Calendar software has come a very long way since the launch of Microsoft Outlook in July 2012. Many calendar apps have features that can help you make the most of your time. Whichever app you choose, look for the system to be feature-rich and able to integrate with your email and other apps.

BEST PRACTICES

Use your time more efficiently with calendar apps that add more automation to your day. Use them to remove tedious tasks that eat away at your time.

Google Calendar

If you are already using a personal or business Gmail service, the Google Calendar is going to be a safe bet for a versatile online calendar. Since many small businesses use Gmail and Google Workspace, the Google calendar is an easy system to sync with the suite of Google tools that the company is already using. The calendar is surprisingly feature-rich—surprising in that many people who have used the application for years have not utilized many of its features.

For starters, Google Calendar can sync with iCal and Outlook, so using one does not mean that you have to sacrifice the other. It makes transitioning to the cloud-based Google calendar easy. You can also easily share your Google calendar with your co-workers, which makes for efficient scheduling of group meetings via Google Meet or a conference call. It also lets others see when you may be unavailable if they are trying to get in touch with you to collaborate. You can set the calendar to show your working hours so that colleagues will know when you will not be available. Again, this is a helpful tool for collaborations and communications.

It also adds in your contacts to increase the ease of scheduling and will send you an itinerary for your day each morning. There are many extensions that can be added to Google Calendar that make your work more efficient and increase your productivity. For example, regardless of what app you use for video meetings, there is probably a Google Calendar add-on for it. This helps you to automate your scheduling by adding in the web address and any needed codes to the calendar invites that you send out for your upcoming video conferences. If you use a CRM system to track customer interactions, many of them have an add-on extension for Google Calendar to help you track communications and topics of your meetings.

Calandly

Trying to get several busy people to find a time to schedule a meeting or event that fits into everyone's calendar can be frustrating and time consuming. It can result in numerous emails and calls going back and forth that end up killing your productivity. Calandly was developed to stop that unnecessary back-and-forth to simplify scheduling.

To further enhance your productive day, Calandly can set buffers so that meetings don't run into each other should some run longer than anticipated. This also helps gives you a cushion for some prep time in between meetings. If you struggle with having too many meetings scheduled in a single day, you can add meeting caps so that you limit the number of meetings in any given day.

You can also add team pages for specific groups in your company. This is helpful for outside vendors or clients who are collaborating on the same project. It allows for invitees to meet with multiple team members at the same time and lets project managers set up meetings in a round-robin format to help maintain the workflow.

Woven

Woven is a calendar with a host of features designed for the busy professional. It aids in the ease of setting in-person meetings, conference calls, and video meetings. It adds in automation by allowing you to create templates of the event types you hold most frequently. For example, if the sales team regularly schedules an introductory meeting for new prospective clients or if the design team onboards new clients in a similar fashion each time, the team can set up a template for these recurring events. Even if they do not occur at regular intervals or with the same people every time, it allows you to create customizable templates to reduce time wasted on repetitive tasks.

There is additional automation in the group polls for scheduling. You can set up a poll to ask when invitees may be available for a meeting, and the system will send you that info so you can determine the best time to meet. In addition, the system will analyze your overall time usage to show you where your time is being spent so you can adjust your schedules if you feel the need. It also will suggest possible times to meet based on your team members' calendars and what they have open or already booked.

Reclaim

Reclaim is a tool to help you keep your work life and personal life well-balanced by defending any time that you use regularly. The system analyzes your routines, sets them as habits, and helps you keep those habits. In this case, these are good habits to keep. Let's say that you have a regular morning workout from 7:30 to 8:30 each weekday. You can see the workout on your personal calendar, but on your shared professional calendar, you will simply be shown as unavailable to maintain your privacy.

The system also allows you to rate the importance of meetings and events. If you would like to find more time during the week to work on a specific project, its AI can help you find more time to block out by analyzing your priorities. The system can report on your team members' time and see where it may be better spent. Should it see a project of high priority and notice that a team member is spending too little time on that due to being stuck in marketing meetings, it will help reprioritize their scheduling.

Clockwise

Clockwise is a calendar app that helps you find more time to focus on work. You may be plagued by sporadic meetings that happen at odd intervals throughout the day and week. That will make it difficult to have a large block of time for focusing on one project or a single aspect of a project. Clockwise allows you to go from a fragmented schedule to one with longer focus times to help improve the quality of your work.

The system allows teams to see each other's availability to collaborate. It auto-updates the team calendar from each individual's calendar. It also will analyze the team's schedules and make recommendations to help increase focus times. Clockwise syncs with Slack to let others know times when a user is unavailable. This helps eliminate interruptions and unnecessary communications during periods of deeper focus.

Lightpad

Lightpad is an innovative calendar that focuses on visual stimuli. It helps motivate the user through a linear layout and tags that are designed to stimulate your visual cortex to better plan your week. Most calendars have a tabular layout and are not set up to interact with your perception of time or how your brain perceives the way time flows.

Lightpad is geared toward visual thinkers, and the design is remarkable. The days are in motion in a string of dates that continue in a cascade across the screen. It is set up to make planning a more natural process and make the calendar less cluttered. This allows you to get a better understanding of upcoming deadlines and events at a glance so that you can better visualize the events you have coming up soon and keep better track mentally.

Automation

The automation software market is surging. Undoubtedly, one way to better manage your time is to add more automation to your life. Getting rid of repetitive tasks gives you back small bits of your time. The minutes spent doing things over and over add up quickly. Regaining some of that time gives you more time to dedicate to deeper and more focused work. There is a growing number of apps available to help the busy professional do just that.

BEST PRACTICES

Automation moves repetitive tasks from individuals. Taking away these small tasks can add up to huge chunks of time that you gain back each month.

Zapier

Zapier is a productivity tool that helps you to automate your workflow. It streamlines your processes and eliminates the busywork of daily repetitive tasks. The system allows for you to take separate third-party apps and help them work better together to automate more of your tasks. The company calls these automations *zaps*. They can save a company a great deal of time. For example, let's say you were sent an email from a client with a PDF that contained info on a new marketing campaign. You could create a zap that automatically uploads that document to the appropriate folder in your cloud storage, and then the zap will use a communications app to let the marketing team know that there is a new document ready to be viewed.

In the past, linking email, file storage, and a messaging app from different providers to create an automated task would have required a company a to do a lot of custom coding. However, Zapier does this with no extra coding needed. A user can create a whole new zap from the ground up or choose from existing zap templates that will help add in automation to your routine in just a few minutes.

IFTTT

IFTTT stands for "if this, then that." The service allows you to take an action from one app and automatically trigger an action in another app. You can even take actions from multiple apps to create a single workflow called an applet or recipe. Although the service has mostly been known as a tool for consumers, it has a platform for businesses as well.

Businesses can use the set of IFTTT APIs called Connect to create workflows from mobile applications and web services to form connections and create different automations. They even offer services based on industry to help accommodate more customized needs. Here is a look at some beneficial examples: Let's say you use Mailchimp for email marketing. You could use IFTTT to grab the stats of your latest campaign from Mailchimp, then automatically add those stats to a Google spreadsheet and save the new document to a folder on your Google Drive account that contains your email campaign stats for later review and comparison. You would then have all of your email campaign stats available in a shared folder for anyone on your team to use. Or if your company manages your social media, you can use a recipe to automatically post to Facebook anything that you post to Twitter. You can go a step farther to have any pictures that you post on your company Instagram page be automatically downloaded and stored in a Dropbox folder to keep track of the media that you've used.

Tray.io

Tray.io is an automation platform that uses a visual drag-and-drop editor to create automated workflows. The advantage is that you can see what your automation will do as you put it together. Its automation and integration platform uses connectors to join apps. It has a universal connector and some that are specific to some of the most popular business apps available. You can create workflows between just two apps or very robust integrations from multiple services. Create workflows by department or workflows that help individual departments within your company, such as sales and marketing, to enable them to work better together.

DocuSign

Although DocuSign is not a dedicated workflow automation system, it does help with automation in a very specific arena. It helps automate getting documents distributed and signed to speed up your workflow. It's a tedious task to go through the steps of printing an agreement, signing it, scanning the executed document, then emailing it so that the person on the other end can repeat the same steps. DocuSign eliminates that cycle and turns getting documents signed and distributed into an automated and easy task.

When you get set up with DocuSign, your business processes will be slimmed down and standardized. This allows for you to increase the speed that you do business by increasing accuracy and allowing the automation to take over. The services go far beyond simple electronic signatures. Should you need a document signed by more than one person, the system will send the individuals an email, link each to the document, bring up the areas where the individual person needs to sign, move on to the next section they may need to fill out, and then complete the process by sending all necessary parties copies and alerting them when all parties have completed their parts. It can also help through the entire life cycle of an agreement by allowing for the tracking of any changes made, maintaining multiple versions, and keeping a library of contracts to help speed up provisions and amendments.

DocuSign also boasts some impressive security. This helps maintain compliance with regulations such as HIPAA and FedRAMP. The system also integrates with a number of third-party apps, including leading CRM services, which further assists in your process automation and speeding up the workflow.

Hootsuite

Hootsuite is a service to help manage and automate your company's social media presence. It can help you keep your brand messaging and image consistent across all social media platforms. You can schedule posts, curate content that is valuable to your followers, and quickly produce relevant analytics. You can send out posts in automated batches so that they conform with the time zones of your audience and have better impact. All of this helps save valuable time and also gives you a marketing edge against competitors.

Hootsuite is one of the more mature social media management platforms available. It is feature-rich and has many help documents and tips. Its dashboard is user-friendly, which allows for a new user to get started quickly and hassle free.

Two-Way and Group Communication

Meetings can be productive, but too many meetings can be a waste time. Even the endless chats and emails can add to the time management issues that people experience when the team is distributed. We still need to communicate and express ideas to each other, but most times, there are better ways to do it than a meeting. Tech companies have heard the call of those stricken by meeting fatigue and have stepped up with some new innovations to help you get your message across and keep your time freed up.

BEST PRACTICES

Look at alternative ways to communicate with remote teams that cut down on the time spent messaging back and forth.

Loom

Loom allows you to cut down on virtual meetings by quickly creating a video that can capture your computer screen and use its camera and mic and create a message. You can use this to respond to people instead of typing out a long and complicated email. It could be a support video that you can send to customers for frequently asked questions or a quick and inspiring video clip for your team to explain an aspect of a new project.

Instead of scheduling a meeting, you can automate by sending a Loom video. This eliminates scheduling issues and makes sure that nobody misses part of what you are trying to convey by allowing them to replay any parts they didn't catch the first time. Use it to automate internal onboarding processes or for training purposes or to showcase new features that your company wants to announce to clients.

Yac

Yac is leading the way in asynchronous team communications. In the always-on collaboration platforms, you can get pulled into lengthy chats and drawn into message rooms. These will not only pull your focus away from your work, it creates multiple streams of interruptions. Yac is designed to remove those productivity nuisances that are inherent in remote team communication. By using asynchronous voice messaging, team members can get the info they need, express themselves better, and not feel the pressure to do any of it in real time.

Think of it as a next-gen and more purpose-driven voicemail. Yet Yac is still far more than that. Let's say you want to share a screen shot of some web analytics with your marketing team. You can share the image with your voice message and ask for feedback. You can then see who has heard your message and seen the image or just wait for the feedback to come in for you to hear. This gets rid of the constant back-and-forth that occurs when people add in info that comes in real-time group communications. It cuts down on interruption and lets team members better manage their work priorities and their time.

Pragli

Pragli seeks to remove the loneliness that comes with remote work and uses avatars designed to resemble coworkers to help you feel like you are all together. It's a format that empowers teams to eliminate the need to schedule a lot of video meetings.

The avatars are ever present and will indicate if the team member is focusing on a project or in a meeting or done working for the day. You can then choose if you want to just send a chat, wait unit they are done with a task, or even interrupt if its urgent.

One feature I love is the virtual office hours. When the team is distributed, getting an idea across can take several virtual meetings or multiple emails. Dedicated office hours allow for teams or key individuals to work together virtually. You host your room where you are working in the virtual office and your teammates can come in to join you in that room. It mimics the feel of when you were in the office in person and they would come to your office to ask a question or get clarification on a project. At the end of the day, Pragli can help break down the barriers that separate a distributed team by making room for more self-expression and informal collaboration. This helps save time and enhances virtual communications to make it feel more like in-person communication.

Sococo

Sococo takes the virtual office to another level, offering a mixed reality perspective. You can create a virtual office setting where everyone in the company comes in for work each day. It will show the lobbies, various conference rooms, break rooms, focus rooms, and more. It will also show which team members are in what rooms, along with their current activities. You may see three members of the marketing team in one conference room and want to chime in about the project they are discussing. If so, you can knock and ask to join in. You may see others taking a few minutes of downtime and socializing in the break room and go over there to chat or catch up.

By recreating your office in Sococo, you allow for impromptu and more real-world collaborations. The design has the intention of bringing distributed teams closer together to have better quality remote collaboration and less wasted time.

Distraction Blocking

Online distractions are a growing cause of time management issues. When you think about it, how much time do you spend on one screen or another? We watch shows on demand on one screen, work from another, video chat and message with friends from another. With so much of our personal and professional lives spent on a screen, we tend to conjoin different aspects of our lives on multiple screens. We don't really dedicate one screen for work and another for play. They intermingle and many times throughout the day, personal items will pop up on a work screen distracting us from the job at hand.

When it happens too often, we find ourselves losing track of how much time we spend focused on these distractions. For many of us, this causes us to end up feeling frustrated that we don't have enough time in the day to get everything done.

In the workplace, people tend to have the same distractions. As Figure 5.1 shows, many workplace distractions also apply to working remotely from home. In fact, you can get a double dose of distractions while working from home. The co-workers may not be in the same building, but they will still reach out via chat, phone, and video drop-ins. Then you have your friends and family, deliveries, and other interruptions.

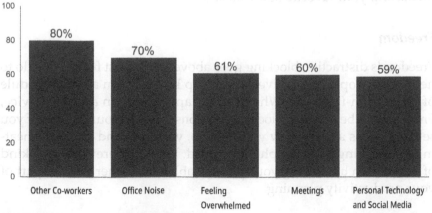

Figure 5.1: Top workplace distractions

When a crisis hits that will cause your company to work remotely, everyone will experience a level of anxiety. As team members continue to be separated, that anxiety will be ongoing. During times of stress, it is natural to find yourself and your team seeking comfort in more recreational activities that keep you happy. When you are in front of your work screen, sometimes you end up surfing on social media sites, doing some online shopping, or socializing with friends and family outside of work. That's part of the human equation, and you should give yourself and teammates some leeway during your crisis when it comes to these distractions. But it's easy to let things slip to the point where they get a bit out of hand. That's where some savvy time tracker apps can help guide you back to get a better command of your day.

RescueTime

RescueTime helps busy people discover where their time went during the day. It works quietly in the background and tracks the apps and websites that you use during the week. Many times, you will find that you bounce back and forth between the same handful of pages in short increments. However, those small increments add up quickly into a large amount of time by the end of the day and week. RescueTime helps you discover these habits and analyze how you spend your time. This information helps you become more conscious of how often you visit websites that are time wasters for your work day.

Beyond just tracking websites and apps, you can use the service to also track time spent on meetings, calls, and focused on work, the goal being for you to have longer stretches of time for deeper and better focused work. Setting up goals and alerts can help keep you on track and freer from distractions. You can also add settings to block specific websites during work hours so that you can further avoid the temptation of social media and your favorite news sites.

Freedom

Freedom's distraction blocking goes above what most focus apps do to help keep people productive. It can help keep you on a tight schedule of distraction-free time. While similar apps work on a single device, Freedom can be set up to block distractions on all of your devices. If you set up a focus app on only your laptop, you may find yourself cheating by reaching for your phone instead. Freedom prevents that kind of temptation by making your phone, tablet. and other devices part of your productivity planning.

You can set up advanced scheduling so that you can automate your periods of focus for each day of the week. Freedom allows you to block websites or apps or both. For times of deeper focus, you can even block the Internet altogether to keep your mind distraction free. For times when you really need to keep your commitment to a period of highly focused work, you can even set up the locked mode. That will help you keep your commitment even if you lack the personal willpower to do it on your own. That setting will keep your devices locked out of the list of apps and websites that you have chosen until the Freedom session that you set up is over.

Forest

For many of us, our mobile phones are the source of numerous distractions. Besides calls, we get reminded to pick up our phones throughout the workday by alerts from social media, updates from news sites, and of course, texts from loved ones. Forest is an app that helps you maintain your focus by encouraging you to stop picking up your phone entirely. It even adds a little more enticement for you to keep your phone down by partnering with a top ranked charity, Trees for the Future.

The concept is pretty simple and based on the Pomodoro Technique. Productivity expert Francesco Cirillo created the Pomodoro Technique in the 1980s. The idea is that you break your work down into tasks. You set a timer for 25 minutes and then work on a task for those 25 minutes straight without any breaks or distraction. With Forrest, simply pick up your phone, open the app, and say that you want to plant a tree. Once you do, you will see your virtual tree start to grow. If you make it to the end of the 25 minutes without picking up your phone, your tree will live. However, if you pick up your phone and try to use it for something else, your virtual tree will die. The more you use it, the more trees you will plant, and soon you can have your own virtual forest. Your efforts can also have an impact on the real world. When you use the app, you will earn coins that you can spend to have real trees planted in different parts of the world.

Spark

Email is a definite time grabber and a constant source of distraction through the workday. To combat this issue, many companies now create policies to limit the number of internal emails. That may help slow the flood from co-workers, but it does not help against the deluge of emails from the outside world. Spark helps make your inbox useful again by controlling the flow and prioritizing your emails.

You can set your inbox to receive only emails from key people during set hours. Once those hours expire, the other emails will then hit your inbox. This email snooze will keep the clutter out and let you maintain your focus on the work that needs to get done.

Getting emails is only one part of your inbox dilemma. Responding to them is another. With Spark, you can set aside a specific block of time to write your outgoing emails for the morning, afternoon, or the whole day. Then you schedule the emails to be sent at the optimal time for each recipient. Scheduling for optimal times is great for professionals who communicate regularly with people on the opposite coast and in other time zones. You can also add in some automation to keep up your productivity by creating email templates for FAQs and most frequently used responses as well as set up alerts to follow up on specific emails.

Time Tracking Software

Many professions require that employees track time for the sake of client billing. For example, law offices, ad agencies, accountants, consultants, and others keep a careful eye on the time they spend on projects so that they are accountable for the billable hours on each invoice. However, other businesses can also benefit from tracking because it shows how time is being used up on different tasks.

By looking at this type of data and seeing where you use your time and where your teammates use their time, you can analyze where resources and workloads can be reallocated to increase efficiency and productivity. The insights can help you adjust target dates, get a better sense of where you are on mid- and long-term goals, and thereby help you make more informed business decisions. For small businesses and freelancers, the reports generated from time tracking apps can be a goldmine of revelations on where the hours of the day go and how you can tweak habits to keep your business thriving.

Time tracking software should not be confused with employee monitoring software. They serve different purposes. I will discuss employee monitoring software in more detail later in this chapter.

Toggl Track

Toggl Track is a simple-to-use and highly intuitive app that acts as a very robust timer. You can tag sections of time for projects or clients and specific work for clients. Starting a session is as easy as typing in what you are doing that session and clicking a button to start the timer. Then you tag the session by client as well as the type of work being done. You click the timer again when done working to stop it, then the info gets loaded onto the dashboard and is instantly usable for sharable reporting.

Toggl Track can be download to desktop or mobile devices and integrated into browsers. This interoperability allows for a high level of accuracy no matter if you are on the road, working from home, or in the office. This also makes it a great app if you are unable to get into the office because you have to work remotely. The system can work when you are offline, so if you lose Internet access, you don't lose the ability to track time. Toggl Track integrates with a number of commonly used business apps such as calendars, CRM tools, and project management tools. Setup is designed to be simple, so adding Toggl Track to your suite of time management tools is easy to do quickly.

Everhour

Everhour is another easy system that can work for one person or your whole team. It has strong team tracking insights so that you can easily see how everyone is doing. With a glance, you can see on a timesheet if one person is spending an excessive amount of time on a particular task for a client project. You can then step in before they are overworked and on the verge of burnout and send them more resources and more help to keep the project on track and the team healthy.

Everhour is designed to work well with many of the major project management software systems available as well as with accounting apps. Its native integration puts the time tracking controls right into the other apps to ease your ability to track time. This allows you to easily set budgets, invoice clients, and avoid switching between apps to log time appropriately. The system automatically syncs with each team member's timesheets so you get instant access to the most up-to-date data available.

Harvest

If you are looking for a time tracking tool that integrates with just about anything you may already be using for your company, then you may want to take a good look at Harvest. It can keep you and your team organized and ensure that time is tracked as it should be. The system can be downloaded onto a desktop or mobile devices as well into browsers. It continues to work even when you are working offline, making it an ever-present tool. If there is an app that is not yet natively integrated with Harvest, you can still use tools such as Zapier to connect the software and add more automation.

Harvest can help you track expenses and create invoices, which helps makes things easier for your accounting team. It quickly generates clear reports that help you make sure your team's workload is where it should be and that projects are on track. However, getting started using the features is not as quick as some of its competitors. You will want to block some time aside to get the system set up and customized the way you want it to be.

TMetric

TMetric is a user-friendly time tracking platform that can help make sure your company invoices correctly when different team members charge at variable pay rates. By monitoring activity levels, you can gauge productivity on projects and adjust targets with workloads as needed.

Reporting can be brief or very detailed, with projects broken down by tasks so that you can automate your billable rates by the type of work that was done. TMetric works on desktops, mobile devices, and browsers to help make sure all work is tracked properly for you to invoice, ensuring that you get paid for every minute of your billable time.

Monitoring Remote Workers

Employee surveillance is not a new topic, but it is one that has been receiving more attention now that companies across the globe have been forced to have employees working remotely during the COVID-19 pandemic. Many employers believe that people need to be seen in order to be sure that they are actually working. When pushed to allow the team to work from home, employers and managers wonder how they will measure productivity if the team is out of their sight.

On March 11, 2020, the World Health Organization declared COVID-19 a pandemic. Later that same month, Google searches for the term "employee monitoring" spiked to an all-time high. In May 2020, stay-at-home orders had come into play, forcing companies and employees to adjust to their new work-from-home reality. During this time, news outlet CNBC reported on the surge in the worker monitoring industry. In particular, the report stated that employee productivity monitoring company Prodoscore had seen a 600 percent increase in interest in its monitoring services. This should tell you that the potential for productivity loss due to staff working remotely is not the issue. The employer's *fear* of productivity loss in the work-from-home (WFH) environment is the real issue.

Fear of Loss Overrides Trust

As we have discussed, when a crisis hits that causes your company to start working remotely, professional lives and personal lives collide. That is also true of privacy when working from home. There must be a delineation of monitoring work and respecting personal privacy in order to maintain a level of trust. How management monitors employees' work and productivity is also of high concern. This is especially true if the employee is working on their own computer from their home. For small businesses, a bring your own device (BYOD) workplace is not uncommon. Many people would find an employer monitoring them on their own personal device a breach of privacy. Employers should tread carefully and with a good deal of transparency should they choose to monitor their remote workers.

One of the revelations that came with the sudden work upheaval of the COIVD-19 work-from-home (WFH) orders was that employers became hungry for knowledge about their newly remote workers. They wanted to know not only if employees were at their computers during work hours, but also what they were doing during those work hours.

Employee monitoring services were eager to fill that need and offer up data. But there is an argument that this type of surveillance creates more of a transactional relationship with the employee and erodes the relationship of trust. Many would further argue that a person should expect privacy within their own home, regardless of whether they are on the clock or not. This makes it difficult to determine how far is too far when it comes to watching what remote employees do during work hours.

Ramifications of Monitoring Workers

Privacy laws vary from country to country, and employers should be cautious as to how far they go when it comes to remotely monitoring their teams. In August 2020, the United Kingdom's privacy watchdog, the Information Commissioners Office (ICO), was launching an investigation into the British multinational investment bank Barclays. The formal probe was launched in light of allegations that the financial company was spying on staff members by using software monitoring how they spent time at work.

"People expect that they can keep their personal lives private and that they are also entitled to a degree of privacy in the workplace," an ICO spokesman said. "If organizations wish to monitor their employees, they should be clear about its purpose and that it brings real benefits. Organizations also need to make employees aware of the nature, extent, and reasons for any monitoring." This is not the first time that Barclays had come under fire for such monitoring tactics. In August 2017, the financial institution was exposed for implementing a system that used heat and motion sensors to monitor how long employees stayed at their desks.

In the European Union (EU), the General Data Protection Regulation (GDPR) helps ensure that organizations protect and are held accountable for the information they collect. They must inform employees of the method they use to collect data on them. They must also gain consent prior to collecting data and protect the privacy of the data collected. The GDPR applies to all organizations within the EU as well as organizations outside of the EU that have employees who are within the borders of the European Union.

In the United States, the Electronic Communications Privacy Act (ECPA) is a federal law that is meant to protect the privacy of individuals and their communications, even in the workplace. It is an update of Federal Wiretap Act, which was limited to oral and wire communication. The ECPA covers a much broader scope of electronic communication, including email and digital recordings. This law prohibits employers from eavesdropping in on employees' communications. However, there are two big exceptions to this rule that favor the employer.

One of these is the consent exception. Plainly stated, if the employee agrees to being monitored, then it's okay to do so. But even with this, there is a gray area. Often, when a new employee is being onboarded, that consent is buried in a small mountain of paperwork and the individual is not fully aware of giving permission to be monitored in varying degrees. It is possible that they believe that *monitored* simply means that security cameras are in place in the building and not that their personal emails may be monitored. The consent exemption is not limited to only electronic communications of a business nature. It can include personal communications as well.

The second is the Business Purpose Exemption. This exemption allows for an employer to monitor oral and electronic communications if the company can show a legitimate business purpose for such monitoring. This wording is so broad in scope that a business could use any business reason to justify its monitoring of employees, from being proactive against company theft of time to productivity measurement.

Some states in the United States have added layers of legal protections for the workforce. In Connecticut and Delaware, the employers must inform their employees regarding monitoring communications prior to any surveillance. In California, two-party consent laws require an employer to get permission from both parties prior to monitoring or recording calls and conversations. In Maryland, before a call is recorded, all participants must give consent. So if there was a conference call with five people involved, all five participants must agree to be recorded. But just because something is legal does not necessarily mean that it is ethical. An employer needs to think about how such policies affect the company brand, the workers' morale, and the company culture. Those are questions that affect long-term company goals.

Nobody Wants to Work for Big Brother

Even if you put verbiage in employee onboarding documents and an employee handbook that their work will be monitored, there are still some ethical considerations. In many cases, the law allows for a broad interpretation of what being monitored means, and it may not be in line with what the employees expect. Remember, if you have ambiguous language that there will be monitoring, it is possible that employees believe this to simply mean that security cameras are in place in the building and not that their computer cameras may be included in that consent.

There are a number of employee monitoring features embedded within productivity programs that could be construed as an invasion of privacy. For example, some will take screen shots of an employee's computer either at set times or at random intervals. If the worker is operating on a company-owned computer, that may be expected. If the person is using their own computer to work remotely, they may feel that their privacy is being infringed on.

Some popular employee monitoring services will go as far as using the web cam of the device that the workers use to take a picture at varying intervals during the workday. If that method is used and the person is working remotely in their own bedroom when the picture is captured, they may feel that their privacy is being violated. Take into further consideration that not all remote workers dress fully during their workday. There are many stories of people only dressing from the waist up for online meetings when working from home. To capture and save a picture of a coworker at home in their undergarments, even inadvertently, may be considered a form of misconduct, regardless of whether it is within the rights of the employer or not.

Transparency Is the Best Strategy

There is something to be said about a reasonable expectation of privacy. The idea of what is a reasonable expectation is up for debate, which is why employers should be cautious in their monitoring tactics. Let's take a look at the attention tracking feature formally available on Zoom.

As discussed, the use of the Zoom video conferencing app skyrocketed in the beginning stages of the pandemic lockdowns of 2020. The company ramped up features in order to capture a larger market share. As the company was making headlines, one of the existing features that was brought into the spotlight was the participant attention tracking feature. Although a blog post from Zoom announced that the feature was geared toward educators, it was not limited to the education industry. The feature allowed the meeting host to enable an alert if a participant did not have the Zoom app as their active window. In other words, if the attendee minimized the Zoom app and was on another website or on social media for more than 30 seconds, the host would be alerted.

This feature, along with a number of other privacy issues with Zoom caused a public backlash against the company in March 2020. On April 1, 2020, the company removed the feature altogether. Many would argue that since people were aware of the feature, its capabilities were not causing any privacy issues. However, when compounded with other privacy concerns that the company faced, it was better for the brand image to cease such monitoring altogether.

This should reveal that it is the overall strategy of monitoring and the tactics used that come into question. What do such methods say about the company that uses them and what does it say about the management that supports the tactics? I am a firm believer in hiring the right people and giving them the space to do their jobs. If you cannot trust them to do their jobs, perhaps they are not the right candidates. Or, perhaps, you should not be a direct manager of the team, and you should use your energy more on the client side of the business and not day-to-day operations.

Should you choose to use employee monitoring software, be fully transparent about it. Unless you suspect some form of malpractice or theft, there is no good reason to monitor employees without their full knowledge. Employees would consider that type of behavior to be surveillance, not monitoring. Instead, have employee policies that include forms that expressly detail how you are monitoring, if you are storing that info, and who has access to that data. Any healthy relationship is built on trust. If you can't trust your employees, they can't trust you. If there is a lack of trust, it does not speak well of your brand, how the company is operated, or the company culture that you are fostering.

Company Culture and Trust

In October 2020, retail giant H&M was fined $41 million by a German privacy watchdog for violating the privacy of its employees. The data protection commissioner in Hamburg, Germany, reported that the Swedish clothing retailer had collected and stored personal data on employees with information regarding their health, religious beliefs, family gatherings, and more. This data was stored on a hard drive accessible by over 50 company managers. The commissioner stated that the network hard drive was "used, among other things, to obtain a detailed profile of employees for measures and decisions regarding their employment." The statement continued to say that the hard drive contained 60 gigabytes of data and revealed that superiors at the H&M site in Nuremberg kept "detailed and systematic" records about employees' health, from bladder weakness to cancer, and about their private lives, such as family disputes or holiday experiences.

There should be a healthy balance between the company's business interests and the employee's personal privacy interests. In the case of H&M, it is clear where the company crossed the line and intruded into the privacy of its employees. The internal repercussions of such a discovery can be disastrous for any company. An invasion of privacy in such a manner destroys any trust that employees may have had in the company. The management will be viewed with disdain and contempt. Someone who is unhappy with their job will not do their best work and will not be productive. This also has a serious effect on employee retention. Actions such as those taken by H&M completely negated the purpose that many companies have for monitoring employees in the first place: to improve productivity.

Put aside the legal ramifications for a moment and look at the impact on your brand and the whole of the company before you employ any worker productivity monitoring strategies. Decide what is the company culture that you are trying to encourage in your team and their relationship to the business. If you want to empower your people to do the best they can for the company, then you have to do the best you can for them. That begins with building a culture of trust. As you can see from Figure 5.2, the overwhelming number of business leaders choose trust over tracking software when it comes to employee productivity.

Measuring Productivity, Trust vs. Track

Most Leaders Favor Trust over Productivity Trackers

78% Prefer trust over tracking tools

34% Use tech-based tracking tools

29% Require employees to report how they spend their time

19% Can't get a sense of their team's productivity

Figure 5.2: Measuring productivity

There are good business reasons for monitoring activity to improve productivity, as shown in some examples previously discussed in this chapter. If you see that some team members are taking longer to accomplish their tasks, you may see that the workload is unbalanced and can then allocate more resources to those team members. Doing so limits the amount and methods of monitoring so that its purpose is to help the team. Once you go beyond that, you are crossing the line into surveillance for the sake of the company's bottom line instead of the sake of its people. One helps foster a culture of care and empathy, while the other creates a culture of distrust where people become clock watchers who are disengaged with the company and long for better opportunities elsewhere.

Monitoring Software

It is important to weigh the benefit of any technology that you may employ for monitoring. Have a strong rationale for using it, and allow room for people to voice concerns over its potential misuse. Be open and empathetic to those concerns. That empathy and openness creates a stronger bond of trust, which in turn encourages stronger loyalty to the company and better quality of work from the team. They want to help build something that they believe in. They want to work for a company where they feel valued, trusted, and welcomed.

Any surveillance that has to be done regularly in secret is not good for your company culture. Do not use it as *tattleware*, as it has been dubbed by some in the workforce. Be fully transparent in any monitoring software you use. Let employees know how it will be used, when it is being used, and how it will help them and the team as a whole. Make sure that monitoring practices align with your objectives of improving skill sets, teamwork, and performance in the same manner that a well-rounded employee evaluation would. Make sure it allows for suggestions on ways to improve and also showcase strengths in the team members. And again, ensure that it is a two-way conversation, where employees can feel safe voicing their needs as well as their concerns.

Prodoscore

Prodoscore uses its system to create an employee productivity score much in the same way a FICO score reflects your credit history. By looking at data points through the day, such as phone use, websites visited, use of company online tools, and processes, the system will generate a score that tells managers who is performing at what level.

By reading the scores, a team leader can see who may be less engaged with a project or with the company as a whole. This can be a sign for the leaders to reach out and see what can be done to increase worker satisfaction and their productivity. The productivity score is designed to help increase visibility of the day-to-day activities of team members and remove any doubt of whether work is getting done.

Hubstaff

Hubstaff is a well-rounded service that provides tools such as project management, time tracking, monitoring, payroll, and invoicing. The employee monitoring features run fairly deep. The system will monitor keystrokes on devices to see how much work is getting done at various times of the day. There are options to take screen shots at set times or variable times.

GPS and geofencing are also rolled into the Hubstaff arsenal of tracking features. With this, team members can track time spent at job sites, travel time, and stops made along the way to keep accurate track of billable time. This data can easily be pulled into team and individual timesheets.

Time Doctor

Time Doctor focuses on time tracking and budgeting, making it a great tool if you have independent contractors as part of your team. You can forecast budgets, make payments, send invoices, and check expenses.

Like other monitoring software, Time Doctor can monitor web usage, but it can also remind people when they visit non-work-related websites and see who may have been late checking in. It also monitors app use across a multitude of devices, so it covers mobile and desktop devices in productivity reporting.

Time Doctor will also allow you to brand your reporting and share it with your clients to provide them with transparency. This helps clients know that their project is being worked on, that billing is legitimate, and that work is progressing as it should.

Time Managing as a Whole

High stress levels will hinder time management and productivity in even the most seasoned professionals. During a time when you are forced to take your company fully remote, there will undoubtedly be stress and frustration, both in the leadership and in the rest of the team. Be sure to keep your deadlines as solid targets, but don't forget the human factors in your equation when calculating productivity.

The remote work environment during a crisis will have stress that can cause both familiar and unfamiliar distractions to have more power over your focus. The situation will make distractions more alluring as a form of escapism. This is also part of that human factor. If you are aware of it, you can be prepared for it. Avoid the urge to knuckle down into a no-nonsense and stringent form of management. Remember that your employees' worlds are merging and the delineations between their private lives and professional lives are becoming hazy.

As is exemplified in the interview at the end of this chapter with Scott Baradell of Texas-based company Idea Grove, not everyone wants to be separated from the rest of the team in a remote work environment. Prior to being forced to have his team work remotely, he polled his team about voluntarily going remote while their new space was still being remodeled. The answer he received from the team was a firm *no*. Being forced into a situation that they don't want to be in will absolutely be stress-inducing for everyone.

Make sure that you remain empathetic to the fact that some of the team may have more of struggle than others in this sudden upheaval. That is where some of the time tracking and product tracking software discussed in this chapter can be of great use. Leaders would be able to see where some team members are floundering even if they may not feel at ease telling you outright that they are having issues. The software can help present analytics that will bring opportunities for managers to reach out, see where employees could use some help, and make sure they are keeping their team working cohesively.

Part of recognizing those opportunities will come from honing your listening and communication skills during the crisis, which will help keep the teamwork alive and healthy. We go into detail of best practices for managing the remote team and keeping up teamwork in general in the next chapter.

SCOTT BARADELL, CEO IDEA GROVE, LLC

Company Profile

- Location: Dallas, Texas

- Employees: 28

- Primary Line of Business: As a unified agency, we provide both public relations services and digital marketing services to our clients. We help our clients "Grow with TRUST" by providing Third-party validation, Reputation management, User experience, Search authority, and Thought leadership services, which functionally include (T) media, analyst, influencer and customer validation (PR); (R) reputation management across review sites and social media (PR); (U) trust-centered web design (web design); (S) branded/non-branded search (content marketing, HubSpot consulting); and (T) content marketing (content marketing, HubSpot consulting).

- Primary Audience: B2B technology companies and their buyers

About Us

Idea Grove understands the power of trust-centered marketing because we owe our very existence to it. The agency started in 2005 with a popular blog by founder Scott Baradell, who discussed his profession and its challenges with vulnerability and self-deprecating humor. This instilled trust in readers, many of whom ultimately became Idea Grove clients. To this day, the agency has earned nearly all its business organically, through PR, referrals, social media, and search.

Over the past 15 years, Idea Grove has distinguished itself as that rare agency that truly "gets" growth-oriented B2B technology companies and their buyers. Idea Grove's unified PR and marketing service offering has been built from the ground up with the specific challenges of B2B tech in mind, including grasping complex technologies, maintaining momentum through long sales cycles, and influencing the decision, making of both business and IT buyers.

Idea Grove hires, rewards, and promotes our team members based on their adherence to evergreen values: Creativity, Respect, Energy, Accountability, Teamwork, and Empathy. The first letters of these words form the acronym CREATE. This helps us to remember that in order to be successful, we must CREATE *every day*. We believe that when we adhere to these values, Idea Grove becomes more than a job for our employees and more than a vendor for our clients.

What was the trigger that day you went remote so suddenly?

Before the COVID pandemic struck, we already had a flexible work-from-home policy in place. Our employees could work from home two out of the five days in a workweek, or 40 percent of the time. We asked everyone to be in the office on Mondays and Tuesdays and then pick one more day out of the remaining three days to come into the office. The infrastructure was already in place to support this plan. When the COVID pandemic struck, we were smack-dab in the middle of a move from our old offices to our new offices and being temporarily housed at a co-working space while we waited on the construction of the new offices to be completed. Before moving into the co-working space, we even asked everyone in the company whether we could work from home 100 percent while we were finishing up the construction. We got a resounding NO! So, on February 28, 2020, we moved out of our old offices and moved into the co-working space. Then COVID hit in early March, and by March 11 we were 100 percent remote, in spite of that resounding NO!

What apps, services or technology did you use to bridge the gap from in-person to remote in order to keep the workflow alive with your team?

Prior to COVID, everyone had a laptop, and we were already using cloud-based applications, so we didn't have to retool the entire workforce. We use a suite of software applications to help manage the work. The big three applications are Teamwork for hours tracking and task management, Zoom for video calls, and Slack/e-mail for asynchronous communications.

You said that your team already tracked hours for clients. What software did you use for that?

We use Teamwork to track hours against tasks. Because it is a cloud-based app, it was very easy to move the work from the office to the home office.

You said that you were not a fan of companies using employee tracking apps just to gauge productivity. Why is that?

We don't believe in activity tracking. We prefer to lead with trust. If you don't trust your team to get the job done, whether at the office, or outside of the office, why would you hire those team members? In addition, we have a set of clients that would let us know if we were letting them down, so we really have an immediate feedback loop letting us know if we have a problem. Finally, we are implementing open-book management using the Great Game of Business methodology, where all of the employees understand the financials and have a stake in the outcome of the performance of the company. This level of transparency helps keep everyone accountable to everyone else on the team.

Did you notice much difference in how your team worked together when remote?

The biggest impact of working remotely has not really been the efficiency of completing the work, but instead it has been the reduction or elimination of informal and unscheduled interactions that naturally occur when working together in the same space. It is the watercooler talk or the quick brainstorming sessions that are missing. Or the "hey, can you help me think through this" interactions that have gone AWOL. We have taken specific steps to help recreate these in the virtual workspace through scheduled watercooler breaks or having to be more intentional about scheduling a work session to brainstorm or work through a problem.

How did you combat the feeling of isolation for your teams?

Again, we are having to be very intentional about this too. We are doing more virtual Zoom happy hours. We have scheduled Zoom show-and-tell sessions. We have scheduled activity assignments that then get shared in a specific Slack channel. We are using an app called QuizBreaker, which includes weekly trivia about each other, to help stay connected. And we use Officevibe to keep our fingers on the pulse on the team. This is probably the biggest challenge of working remotely, and we are constantly looking for ways to overcome this challenge.

We've also reopened the office, under the published rules of the governor of the State of Texas. And we have invited everyone to come back into the office once they feel comfortable doing so. Out of a company of 28 people, we average four to five people in the office each day.

How did this affect your company culture?

We had a very strong company culture going into the COVID crisis. I would say that the culture was not really impacted much. But we have been very intentional to ensure that it was not impacted, and we have had to do extra to ensure that the culture was not compromised. For example, we've had some personnel changes during this time and have gone overboard with in-depth, cultural interviews to ensure that the new employees were extreme cultural fits to the existing culture. It would have been much harder to bring on employees that were less of a cultural fit.

What would you do differently if you could do it all over again?

I think the biggest thing we would have done differently if we had known that the pandemic was coming was to not sign our new lease and work to take the company 100 percent virtual, 100 percent of the time. We probably would have gotten a membership at a co-working space and only used that space for meetings. With the savings from the rent, we could have purchased every team member a formal workspace for their house and still spent less money than we do on monthly rent. But, looking back, we were very fortunate to have been thinking through a part-time remote work policy long before it was forced upon us. There was a time when the agency was built to be 100 percent work in the office. If we had to pivot from that to 100 percent work from home overnight, it could have been a significant infrastructural and technical challenge.

What words of wisdom and advice could you give to someone else who finds themselves having to suddenly make their operations go remote?

The biggest piece of advice for anyone considering a remote workforce is to start with a culture of trust and transparency. Leading with these two things makes the transition much easier. It eliminates the worry about people slacking off and not getting the work done. But, for most people, this means starting with their own internal beliefs about employees and potentially changing those beliefs in order to get the best possible outcome. This might be the biggest barrier to a successful transition to a fully remote workforce.

Group Tasks: Keeping the Teamwork in Your Team

One of the challenges company leaders found when their company was suddenly forced to work remotely was to keep up the level of teamwork. It's not that the team could not work well enough to meet deadlines. They still were able to accomplish group tasks, yet something was missing. There are intangible elements of working as a team: the camaraderie, the nonverbal communication that close team members can pick up on, and the predictability of each other's actions that colleagues develop when everyone is accustomed to working in person.

It's similar to the way an orchestra performs complex symphonies after weeks of practice together. In that scenario, the conductor is in charge of making sure everyone plays well in concert with each other. If the maestro were conducting the orchestra remotely, they would need to change the methods employed to gain the same effect. In a business setting, team leaders need to change the way they manage their teams if they are to keep up the teamwork and empower team members to work cohesively even though there are miles between them all.

Having a good remote work continuity plan will help a company survive should they be forced out of their place of work. However, in order to thrive, you must look at what hurdles your team will face when they are on their own but still expected to function as a cohesive team.

A remote leadership survey by engineer recruitment company Terminal surveyed 400 HR and engineering leaders during the COVID-19 shutdowns. It found that remote work strategies in a number of companies failed to address what would end up being key issues. For example, 63 percent had plans to address remote productivity, while only 21 percent had plans to address the burnout that employees face from a sudden, forced work-from-home situation (Figure 6.1).

Remote Work Planning

Leaders are focused on short-term productivity, not burnout or loneliness.

63%   Plans that address productivity

32%   Plans that address employee loneliness

21%   Plans that address employee burnout

19%   Only 19% of business leaders report having a remote work strategy in place prior to COVID-19

Figure 6.1: Remote work planning

In this chapter, we discuss tactics and strategies to tweak management styles in the sudden WFH scenarios. By taking what you already know and making some adjustments, you can keep the team working closely and overcome the challenges of morale and remote work fatigue.

Remote Work Fatigue

An interesting phenomenon was observed after weeks of forced working from home due to the COVID-19 pandemic. At first, work-from-home (WFH) fatigue was played off as something imagined, but executives soon came to realize that it was a very real thing. And it had a negative impact on employee morale, productivity, and teamwork.

Working from home came with some positives for many employees, but for others there were pent-up frustrations. It is one thing to voluntarily work from home. When you have choice, you are exercising your free will and you can pop into the office and work next to others when you choose. When a crisis hits and everyone is forced to work apart for an unknown amount of time, it causes an upheaval that you have no choice in and little control over. That realization can color how one responds to the new challenges of forced remote work.

Lowered Work/Life Balance

One of the reasons that becomes obvious for the work-from-home fatigue is the blurring of the lines between home life and work life. People may find that they are quickly overworked and reaching burnout from the new WFH scenario because it is very hard to switch off for the day. Unwinding after work has become a challenge for many. There is no clear agenda for getting off work when you are suddenly working from home. You don't have the ritual of packing up, leaving the building that you work in, and commuting back home. Those daily activities create a clear physical and mental indication that the workday is over. They give you a signal to disconnect from that part of your life until the next workday, and it readies you to reconnect with your home life.

In August 2020, a worker survey conducted by remote job website FlexJobs and Mental Health America showed some interesting data regarding WFH burnout during the pandemic. The survey showed that 70 percent of respondents have experienced burnout from their work, with 40 percent of them stating that the burnout occurred during the pandemic. Additionally, 37 percent responded that they found themselves working longer hours than normal during the WFH experience of the pandemic.

Without that daily routine of going home, remote workers find it hard to switch off for the day. They find that they make themselves available for work-related tasks at all hours of the day and evening. This is another indication of the imbalance of work and life, which can easily create higher levels of stress and a feeling of anxiety.

Mental Health

Mental health is an issue in every country even during the best of times. During times of personal or professional crisis, the levels of stress and anxiety are heightened. If not checked, these elements not only can lead to burnout but can exacerbate pressures that lead to depression in coworkers. The crisis that causes your team to work remotely could be a catalyst for mental health issues. If a team member had struggled with mental health issues prior to your crisis, those issues may seem more acute during times of increased stress.

It is vital for the physical and mental health of the team that leaders are cognizant of these very real issues when they are in the middle of crisis management. Heightened levels of awareness and empathy are required during these times. If you are able, make mental health care accessible in some form. You can opt to perform an anonymous poll or survey among employees to see how everyone is faring. This may help you discover new ways that you can better support the team. Try to gain some insights on what their stress levels may be and identify anyone who may be struggling.

You want to recognize problems before people begin to feel the effects of burnout. The World Health Organization classifies burnout as an occupational phenomenon in its international classification of diseases. It is defined as follows:

> *Burn-out is a syndrome conceptualized as resulting from chronic workplace stress that has not been successfully managed. It is characterized by three dimensions:*
>
> ■ *feelings of energy depletion or exhaustion;*
> ■ *increased mental distance from one's job, or feelings of negativism or cynicism related to one's job; and*
> ■ *reduced professional efficacy.*
>
> *Burn-out refers specifically to phenomena in the occupational context and should not be applied to describe experiences in other areas of life.*

Encouraging preventive self-care and overall wellness is a must if you are to keep your team safe and healthy. However, promoting self-care is not enough. Strong leaders must take firm action by adjusting their management styles and the way they conduct remote business. This will help to empower everyone to be in a good and healthy team environment within their own home. Acting preventatively is key. Be sure to do this for yourself as well. When a team leader begins to show the signs of stress and being overworked, it creates a ripple effect that touches the rest of the team. If you feel as if you are experiencing symptoms of burnout, take a mental health screening. There are a number of free mental health screening tools available online that are anonymous and confidential.

BEST PRACTICES

Employers should take their employees' mental health into consideration just as they would their employees' physical health.

Perform a Reset

Once you get settled leading a team in your office, you create a certain style of management. You take the skills that you have learned over your career and adjust them to that office work environment. You need to become more agile in your leadership to meet the demands of the crisis and the new remote environment. In this situation, your listening skills as a leader are needed far more than your need to be heard.

Making a conscious effort to adapt your in-person management style to be a stronger remote leader will best serve your team in their efforts to work cohesively. By doing so, you can establish a boundary between work and home, reduce employee anxiety, and relieve stress. All of this combined will create a better mental health environment and create a stronger ability for the team to work together. It can also help foster creativity and boost the imaginative processes.

Encourage Boundaries

One of the major contributors to stress during the WFH experiences of the COVID-19 pandemic was the lack of boundaries between work and home. As discussed earlier, losing the routine of going home made it difficult for people to punch out, so to speak, at the end of the workday. People found themselves working longer days and making themselves accessible for work-related communications into the evening. To prevent burnout and work fatigue, employers should strongly encourage boundaries between work and personal life during the forced WFH time periods.

Know When to Log Off

When dedicated people are working on a project, they tend to lose track of time and can end up working very late. Doing so certainly demonstrates a strong work ethic, but it is not part of a healthy work ethic.

Without balancing out work time with time to decompress and to engage in regular enjoyable activities that reduce stress, burnout is an inevitability. It's not an "if," it's a "when." In short, in order to remain healthy, people need a balance between work and play time. This is a difficult concept for many hard workers to accept. I grew up in a household where hard work was considered the sign of an honest and good person and that relaxing or having fun was considered goofing off and was discouraged. Many people have had a similar experience growing up or have worked with supervisors who had a similar outlook on work ethics. The latter involves jobs where breaks are given begrudgingly and only because it is required by law.

Sometimes, people with such a strong work ethic will only log off if told to do so by someone in a higher position in these situations, it's important for the team leader to encourage logging off for nightly downtime. If the team leader simply gives permission or sets a time for close of business for the day, some employees are more likely to stop working.

Turn Off Emails

It's not very effective to walk away from your computer monitor if you are still going to work from the mobile screen on your phone. Encourage people to set collaboration tools, email, and other commutations tools such as chats to show themselves offline or busy during non-work hours. Turn off work notifications, and keep dedicated work devices asleep or off.

Disconnecting helps set the stage for stronger boundaries between work and non-work hours when working from home. It will help keep teams from reaching out to each other when it is not appropriate and encourage people to maximize their downtime for de-stressing activities. It's like the difference between getting a good night's sleep and a bad one. If you get hours of deep and uninterrupted sleep, your body and mind will be recharged and ready for the coming day. However, if your night is peppered with short periods of light slumber interrupted by moments where you can't get back to sleep, then you wake up frustrated, unrested, and hanging on to the stress from the previous day. You can promote periods where teams can more deeply relax and enjoy their downtime by encouraging them to avoid work communications when they are not absolutely needed.

Encourage Breaks

As we discussed in previous chapters, breaks are essential for maintaining your personal resources throughout the day. Once depleted, those resources are difficult to get back. Feeling run down will only exacerbate challenges that one may be working with and can tend to magnify the severity of issues. Encourage breaks within your team. Have them schedule regular breaks if need be. Be sure that these breaks are not wasted by team members using the time to catch up on other kinds of work, whether it be job-related or personal.

Educate your team on the wisdom of taking better breaks. Get them to do something preferably away from their workstation, or at least something that they will find fun and enjoyable. Spending some time outside with fresh air, sun, and a change of scenery can help recharge a person's mental batteries quickly and get creative juices flowing.

BEST PRACTICES

Team leaders should actively encourage breaks and downtime as a preventive measure against burnout and WFH fatigue.

Foster an Environment of Engagement

When people feel engaged, they are more mentally present. They tend to forget other issues they may have and focus on the moment. A person who feels engaged also feels a sense of belonging, which is important for a team member, especially when work seems stressful. It makes them feel wanted and cared for. It makes them feel accepted as a valued member of the team.

In the summer of 2020, a survey of 2,000 remote workers was conducted by OnePoll for CBDistillery regarding how workers were coping during the WFH period of the pandemic. Seven out of 10 workers said that they struggled to maintain a healthy work and life balance. Further, 65 percent said that they were working longer hours than ever before, while 56 percent said that they were more stressed about their job than ever before (Figure 6.2). Increased stress and anxiety represent an unfortunate side effect of being forced to work from home while being separated from your team for an unknown period of time. However, what causes more stress and makes the WFH situation worse is when you feel that your employer does not care about this added weight. Six in 10 respondents feared that their job would be at risk if they didn't go above and beyond by working overtime, and 63 percent agreed that time off is generally discouraged by their employer.

Remote Employee Burnout

Americans working remotely struggle with work / life balance

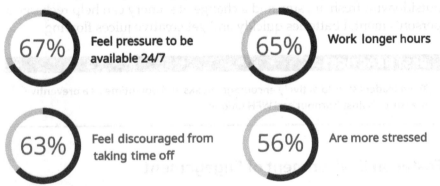

67% Feel pressure to be available 24/7

65% Work longer hours

63% Feel discouraged from taking time off

56% Are more stressed

Figure 6.2: Remote employee burnout

Strive for an environment that values the team as well as the individual, making them feel noticed and appreciated. In the same OnePoll survey, only 21 percent said they were able to have open, productive conversations with HR about solutions to their burnout. Fifty-six percent went so far as to say that their HR departments did not encourage conversations about burnout. Feeling that you have nobody that you can turn to will only hasten burnout. Be sure to recognize your employees' individual contributions. Be open to input. In fact, create opportunities for feedback and input. Schedule time at the end of meetings where feedback can be given. Have an open-door policy where employees feel comfortable coming to you with both concerns and good news.

Establishing an open-door policy when working remotely can be easy to achieve. Simply make your team know that they can come to you to talk at any time about anything. Let them know that you are around by showing yourself online and available in your collaborative tools and communication apps. You can go a step further by showing yourself as "in-office" on your shared team calendars.

If you see signs of burnout in your employees, suggest that they take some paid time off. This will bring back huge rewards for your company, your company culture, and the individuals in your team. That small investment will be paid back many times over with employee satisfaction and retention.

Check in with your team daily. If you see people not being engaged in meetings, check in on them privately. See if they are having some difficulties or stress that may be job-related or from outside of work. Offer a mental health day to those you feel may benefit.

The lines of home and work are already blurred, so encourage time for people to socialize just as they would in the office. Watercooler time is important for team bonding. Allow for time before a meeting for people to catch up and engage with each other on a more social level.

Ask how things are going at home and how people are adjusting to their new situation. If you know about some personal problems, have some individual meetings where you engage that person and ask how things are going. Be sure to check up on them. Be ready to listen and see where you may be able to help (Figure 6.3).

How Leaders Can Support Remote Workers

Employers can help with stress and support mental health.

56% Work flexibility

43% Encouraging time off

43% Offering mental health days

28% Increased PTO

Figure 6.3: How leaders can support remote workers

Expenses for Home Office Needs

A poor work environment can be draining, both mentally and physically. Poor lighting will strain the eyes and can cause headaches. Uncomfortable furniture can cause muscular and skeletal issues and be a source of stress and mental fatigue. Sitting eight or more hours a day glued to a monitor under these conditions will hasten burnout. It creates an unhealthy environment for work and will likely cause a person to wake up dreading the return to work, even if their office is in their own home. It will ultimately negatively impact the quality of their work and increase their stress level.

Invest in your people and that investment will return huge dividends. Get them a better chair, a new lamp, or some noise reduction headphones if their home is too noisy and distracting. Making them comfortable will help keep them productive, reduce their stress, and keep them happier overall. It will also show that you care about their well-being and value them as individuals.

Gratitude

One of the most powerful feelings that can provide encouragement to someone in a high-stress situation is a feeling of gratitude. When I was in my 20s, I used to wait tables in a very trendy restaurant in Fort Lauderdale called Indigo that seemed busy all of the time. During some busy shifts, waitstaff or bussers would not show up, and everything seemed to go wrong. On those nights, I would be sweating from the heat of the outdoors, from going in and out of the kitchen, and from moving quickly for hours without sitting or taking a break. At the end of those shifts, we were all exhausted and highly stressed.

A new manager had taken over on one of these nights, and at the end of the shift when he signed my clock-out slip, he looked me in the eyes and said, "Thank you, Henry." I looked at him oddly and asked him what he was thanking me for. He said for coming in today and doing such a great job. I was flabbergasted. In all of the bars and restaurants in New York and South Florida where I had worked, nobody had ever said thank you to me for the work I had done.

This honest gratitude had a powerful effect on me and the other front of the house staff. When that manager called us on a day off to come in because he was shorthanded, we gladly came in if we were able. I was glad to work harder for him because I truly felt he appreciated me. He was actually thankful for the hard work that I put in during very stressful circumstances, and he recognized all of it. He then established a policy that call-in people were scheduled to come in for every shift to cover the work when others did not show up. So he not only recognized our hard work, he was engaged with what we were doing and recognized the problems we faced. Then he took action to fix them. But most of all, he created an environment of gratitude where there had not been one previously. From that place of gratitude and appreciation, he was able to work with us to solve a good number of morale problems in the restaurant. The high turnover rate for front of the house staff slowed to a trickle and people looked forward to coming in to work.

Never underestimate the power of showing your gratitude to others on your team, especially in highly stressful situations. Gratitude has the power to transform negativity and turn it into positivity. Gratitude is a part of a recognition-rich work environment. It puts strengths first and shows people that they are appreciated and that their efforts are noticed. Feeling that their efforts make a positive difference and are recognized can be the fuel needed to get a person through challenging times.

Collaboration Tools and Usage

Collaboration tools were intended to make projects easier to work on when a team is separated. They help keep items organized, make sure that people keep up on tasks, meet deadlines, and above all, communicate. And people certainly communicated over these various apps. In fact, in many cases, people overcommunicated in an attempt communicate more effectively. Unfortunately, more of something does not equate to better quality.

A higher influx of bad communication just becomes noise after a certain point. When you are forced to listen to too much noise, you become desensitized to those communications. They become unimportant, and even worse, they help drown out your creative voice. In short, you just stop caring.

Being subject to a constant barrage of noise will kill your creativity and your spirit. There is a reason opposing armies will use a tactic of constantly playing obnoxious music within earshot of the enemy. It is a form of psychological warfare, and it is proven to be effective in breaking the will of combatants. This is the exact opposite effect a team leader wants to see in the team, but it's a result that kept occurring during the COVID-19 WFH experiences.

In October 2020, Utah-based collaboration software company Lucid released results of a survey it conducted of concerns from knowledge workers and managers regarding remote work. Knowledge workers are those whose work is based in the knowledge and information of their specific field, as opposed to people who work to produce goods or services. Structural engineers, lawyers, and economists are good examples of knowledge workers. They bring intellectual capital to companies and their industry.

The survey was distributed among 1,000 employed adults in the United States who worked a full-time, traditional desk job from home at least three days a week. The sample included 300 individuals in a management role. Respondents came from enterprise and mid-sized businesses nationwide in all major industry segments. They were equally divided between male and female and included all ages: baby boomers, Generation X, and millennials. The survey showed different perspectives between employees and managers regarding productivity versus creativity. The report showed that creativity overall is suffering because of poor collaboration, with more than a third of managers ranking employee productivity as their biggest concern with employees working from home, while employees report that collaboration with their teams has suffered the most.

The report went on to conclude that creativity suffers further because knowledge workers are distracted when using collaboration solutions. One in four (25 percent) remote workers admitted to spending at least half of a typical virtual brainstorming meeting being distracted, and 62 percent of remote workers admitted to bad behavior while participating in virtual brainstorming meetings from home (1 in 10 have used the bathroom while on a call). When asked what would be most exciting about returning to the office, 37 percent of employees ranked in-person team collaboration as number one, twice as many as the next option (a dedicated workspace without at-home distractions). Nearly one in four remote workers said that working from home has hurt their creativity.

Too Many Video Meetings

Zoom fatigue was a topic that kept getting headlines during the COVID-19 pandemic. At first, this was disregarded as a work-related issue that employees needed to accept as being part of working remotely. But by

May 3, 2020, online searches for "Zoom fatigue" peaked, and employers sat up and took notice.

According to *Psychology Today*, Zoom fatigue describes the tiredness, worry, or burnout associated with overusing virtual platforms of communication. It's ironic that a tool that was designed to help distributed people feel more connected ended up creating a drain on the psyche of workers across the globe. There are well-established sociological and psychological reasons that explain why we feel tired and more stressed from a deluge of video meetings, and we will go into those details in Chapter 7, "Client and Team Meetings: Making the Most of It." In this chapter, we will discuss best practices for using this important technological business tool.

No Back-to-Back Meetings

You should make it a policy to build in breaks between video conferences. If having several video meetings a day leads to stress and fatigue, a marathon of them certainly is worse for your health.

Many calendar apps will help you create time buffers between meetings automatically. Use them. It will give you time to take a few minutes between meetings and refocus. Consider this example: If you ran a race and had a second one scheduled, you would want to take time to refuel and rest before that second one. You can use the same principles with meetings. Schedule time to center yourself so that you can be fresh for the next meeting. This practice will also help you maintain your longevity for the rest of the day and the post-meeting work that still needs to be done.

Schedule a Meeting-Free Day Each Week

Having a day each week for no meetings is one the best ways to keep team members mentally agile and working well. It creates a day where people are empowered to dedicate their energies to deep and well-focused work. This does not mean that they should avoid reaching out to each other to collaborate or clarify needs. This is about keeping the schedule clear of any formal scheduled meetings.

Meeting-free days accomplish several things to keep up the teamwork in the distributed team. For starters, they allow for a better flow of work. People can get into their zone when they know that they will be free of distractions. Many times, I find it difficult to allow myself to get deeply focused if I know that I have a meeting coming up within the next 30 to 60 minutes. I still have to prepare myself for those upcoming calendar events, so deep focus is not an option. A day free of meetings means that teams can have the luxury of several hours in a row to focus more mental energy on completing work. Managers can also get a much-needed break from leading the meetings so that they can have time to knock items off their task lists.

Making a meeting-free day also helps the team retain focus and be more engaged on days that they have meetings. As discussed, too many meetings just create noise. It creates an environment where people feel and become disengaged. Once that seed of disengagement gets planted, it can grow quickly, and people lose interest in their work and the job in general. Reducing the number of video meetings and making meeting-free days helps keep people on track with their work and keeps people from the added stress of being "on" in front of the camera each workday.

Pick Up the Phone

Not every meeting needs to be on camera. Well before the advent of Zoom and Skype, remote workers were able to conduct business across the country and internationally very well using the phone. It was efficient and we were able to communicate with great accuracy to get our message across. I have been in board meetings where members were conferenced in on a Polycom device that sat in the middle of the conference table and we were able to debate ideas, reach policy decisions as a group, and vote on those decisions. Zoom and videoconferencing have certainly been trendy and buzzworthy. Videoconferencing has its place, but it has been overused recently.

Being on camera adds a layer of stress that the team does not need. In some instances, pressure can make some people perform better, and the stress of a deadline can make people be productive in a timely manner. But there are small stresses that are unnecessary, and those are the ones that tend to build up and get people upset.

On a video conference, you can see your team members, they can see you, and you can each see yourself. People sometimes get acutely self-conscious when they see themselves minimized in the meetings. It's easy to fall into a negative space about your appearance and how you carry yourself. If you are already self-conscious about your appearance or how your present to a group of peers, then constant video meetings can add to any negative feelings you may already have about your self-image. This can increase the drain of mental resources and magnify fatigue.

Keep the purpose behind the meeting in mind before you book it. If there is not a visual component needed, then make it a conference call instead. This will help stave off Zoom fatigue and lower daily stresses.

Do You Need the Meeting?

Every professional I know has sat through a meeting that, afterward, they thought could have been an email. According to collaboration software development company Atlassian, the average professional in America spends about 62 hours a month in meetings. That's an average of 15.5 hours in meetings each week per person. That sounds exhausting. In a situation where everyone is forced to work remotely, too many meetings add additional layers of stress. They slowly eat away at the mental bandwidth that individuals in teams need to work cohesively with each other.

Get rid of meetings with vague agendas. Challenge yourself to vet the premise for each meeting to see if it really needs to pull several people away from focusing on other tasks. Doing a cost analysis for a meeting can help put them in perspective. If there are six people in your hour-long meeting, add up the cost the company paid for an hour of their time. Compare that to what a client would have paid for an hour of dedicated time for six of your people to work on a project. That will give you a good idea of what each meeting costs your company. You may come to the realization that these meetings are costing your company way more than you may have guessed and then choose to cut them back to favor more efficient forms of group communications.

Also consider the need for those hour-long meetings. They tend to be far longer than what is actually needed to get the message across. However, they have become a business standard in many companies. More concise and goal-focused meetings help the team stay on track and give them additional focus time for work. You may notice many times that people fidget during hour-long meetings or simply stop paying attention. This is mostly because they have more pressing project tasks that can better benefit from their attention instead of the meeting they feel stuck in. What can be said in an hour-long meeting usually can be said in less than 40 minutes if you keep the meeting on track. Each meeting should have a goal to accomplish, and you should take the shortest path to reaching that goal. Stay on topic and goal-focused with your meetings.

If you keep the team cohesive in how they operate and communicate, then only key people need to attend meetings. The team can trust in their co-workers that the information and the sentiments of the meeting can be relayed properly and in a concise manner. People who are not essential to the meeting can then do more focused work, which enables the team to better function as it should.

Asynchronous Communication

One way to determine if a meeting is unnecessary is to think about whether the message behind it has to be understood in real time. Think about it this way: Did someone really need to drop what they were doing in order to hear what you had to say right then and there, or could it have waited a while? If it could have waited, did the message have to be relayed in person in order to be fully understood? If the answer to both of these questions is no, then an email would likely be just as effective and would save on employee hours.

Make better use of asynchronous communications to empower your team to have more time to do what they do best. Writing a brief in a shared document could replace a meeting. A recorded video can add in any emotional emphasis that you want to leverage in your message. An added advantage to asynchronous communication is that it works equally well for both the extroverts and introverts on your team. In-person meetings can be a bigger drain on introverts who don't want to be in the spotlight or have any extra attention drawn to them. Asynchronous messages level the playing the field and become more inclusive for all personality types.

If a meeting is not collaborative and doesn't require a lot of two-way exchange of information, chances are it doesn't need to be a meeting. Even most communication that does require some form of response can save time by being asynchronous. A poll or survey is a great example. If you need a team status update, you can send a message out to the group and ask each one to respond with their updates before the end of the day. It avoids wasted time with an unneeded meeting, it keeps everyone in the loop, and it allows for people to respond in their own time as their individual schedule allows but with a deadline. Leverage more asynchronous communications over meetings to communicate with your team members in order to give them more space to work and reduce stress.

A Round of Applause

Going back to showing gratitude and recognition, you can demonstrate appreciation for efforts in the beginning or end of your meetings. It's a practice that will go a long way to increase job satisfaction and reduce stress on all fronts. Professionals are motivated by doing their work well, but people also want to feel appreciated. It is a human condition and should be taken into account along with the other human factors in remote teamwork.

Saying thank you in front of the group creates more opportunity for engagement with individuals and the team. Recognizing accomplishments and especially those efforts that go the extra mile is an investment of time that pays huge dividends in the long run. You can even increase that feeling with small gifts of appreciation. Gift cards to local coffee shops or stores allow them to buy themselves an added reward for a job well done and is a nice add-on to voiced recognition. And don't ignore the power of favorite foods. A basket of goodies goes a long way to help keep people smiling every time they enjoy a treat they have earned.

Emphasizing the importance of individuals and their accomplishments can be a key factor in helping people who are isolated by remote work to feel that they are a vital part of the team and that their contributions make a positive difference to their peers. This will give each person the added fuel needed for their self-confidence and enable them to be a better team player.

Results over Schedule

After you do a reset, try to switch your outlook to a results-first mentality when it comes to your teams and their work. It's important to strive for less focus on control and more on engagement. Unless you are running an assembly-line style of business, the time of day when the work is performed is not as important as the quality of that work.

I have worked with a number of developers who were based in different cities around the world, while I was based in Indianapolis and acted as their project leader. I quickly learned to abandon the thought processes behind traditional work times with us all living in different time zones. As long as their work was high quality, consistent, and on time, when they worked had no bearing on the rest of the job.

Flexible timing doesn't work for some jobs, such as help centers and other positions that are tied to specific hours, but for most remote workers, time of day does not apply for anything other than meetings. Even collaborative work can be done asynchronously with few if any issues. Just as the workplace and employee environment have shifted out of their normal method of operations due to managing your crisis, so must your outlook on work and when it is done.

Allow People to Work in the Right Rhythm

Working in different times zones is something that remote teams become accustomed to overcoming. When your office suddenly goes remote due to a crisis, most of the people will still be in the same time zone, but that doesn't mean that they will still work best from home during those same hours. Remember, worlds are colliding. They may have roommates or family at home during normal working hours that make working from home less than ideal. As I said in a previous chapter, not everyone has a wonderful home life. Sometimes, there are people within their household that they actively try to avoid. Getting focused work done in those situations can be extraordinarily stressful.

People with children may also find themselves suddenly less available to work in the early morning or midafternoon. Even with well-established boundaries for work time, a parent may not always be able to focus well on work once the children come home for the day. This will also be true for workers who have children at home who take their classes online during the morning and early afternoon. Sometimes, it will be better for the company and the team to allow them to finish up later in the evening after traditional work hours. As long as this type of flexible schedule works well with the rest of the team, there should be no reason to not allow it. There are few reasons why it would not work, and a good leader can always make adjustments to empower such flexibility. To deny such flexibility would cause stress on the employee for both their work life and their home life. It could also force them into a situation where they have to make a choice between the two.

Being flexible shows people that they are valued and appreciated. Remember that they are having to give up part of their home and their home life to accommodate the company and its crisis needs. They are making concessions and so should the company. Allow them to plan their workday in the manner that produces the best work. The WFH schedule that works well for some may not be best for others. Empower the team to do what's best for them, and trust that they will work well. You don't have to leave it all up to trust alone. Be sure to build in assurances that the work will get done by setting clear expectations for all.

Accountability and Expectations

One of the first steps to building stronger bonds of trust within the team is to set clear guidelines on what is expected from everyone. This includes leadership. The rest of the team should know what they can expect from each other and from management. This helps with transparency and leaves little room for gray areas. As I mentioned in a previous chapter, the company culture will guide an employee on the right thing to do when nobody is watching. It helps set clear expectations.

When you set up remote expectations and accountability, you make sure that everyone is on the same page as they work from home alone, even at odd hours. This goes far beyond deadlines for tasks and projects. Set up your expectations regarding all aspects of the business being done remotely. It is the only way to be fair and equitable when it comes to accountability.

For example, set up clear guidelines when it comes to communication policies. If people are in different time zones, be sure that there is a window of time each day when everyone is available for one-on-one or group discussions to either clarify needs or simply to reach out for guidance. For example, 2:00 p.m. to 4:00 p.m. Eastern time is a window of time that is accessible to people in time zones across the continental United States. You can set that as a block of time when all employees must be available and online during the week to remove the guesswork for team member accessibility. But don't limit that block of time to the employees. Leadership should then also try to have open office hours during these times. This creates a better communication policy and sets clear expectations for communication and accountability once everyone is on the same page.

Setting guidelines should also apply to *how* people communicate. For example, there should be clear guidelines for something as simple as sharing a document. Do you want the originator of the document to email copies to everyone on their team, or do you want it put into a shared file folder? If put into a cloud folder, how is everyone made aware that it was added to the folder and is now available? You will want them to alert the necessary people on the team that it's been uploaded to the folder and ready for use. But is that done in a chat or in an email or in a checked task list on your project management system? Setting up these guidelines will make sure that communications do not fall through the cracks when everyone is working remotely.

Outcomes

It is also important to address how leadership will analyze employee performance during this time. You will need to be transparent in how teams and individuals will be evaluated. Management should do its best to try to use metrics similar to what was in place when you were all working in the traditional office environment. However, the metrics must be adapted to the remote environment. Established norms such as being in and on time with a 9-to-5 schedule should not apply, but being available during established hours should apply. You will have to decide how you determine if an employee is not performing as they should, as well as how you inform that person. You cannot call someone into your office to discuss the issue. You must decide if you should attempt to meet face-to-face in a coffee shop or if you should have the discussion on a video call. These are not things that you want to decide on the spur of the moment when they come up. Have them documented in a continuance plan and discuss them with the team in advance for full transparency when you have to go fully remote. It enables them to do their best work with little guesswork as to how they should be doing it during your crisis. It will alleviate a good deal of stress for individual employees and make them better team players.

Make your remote management style goal-based with outcomes as the priority. Intellectual output is far harder to measure than how many widgets an individual or the team produces. It's even more difficult to measure in the remote environment if you use outdated metrics or systems that simply cannot apply in a forced remote work environment. As the workers must become agile and adaptive, so must the style of management and how performance is evaluated. Outcome-based evaluation is the best methodology during your crisis. If the team is reaching benchmarks and hitting the goals that are set for them, then they are producing.

To maintain teamwork on the remote team, all of the steps and standard procedures for remote work need to be addressed and spelled out if you are to set expectations. Have these expectations typed up on a shared document that is available to everyone. Should there be any question, they can then simply refer to the document. This also ensures that everyone is held accountable equally.

Open Virtual Office Hours

Communication is key in any healthy relationship. That is true within the team as well as between the team members and management. Maintaining strong and effective communication is one of the biggest barriers that remote teams face, and it is an even bigger barrier during times of high stress. Newer employees on the team and introverts may find new remote work policies intimidating and may be afraid to speak up should they have concerns or questions. Management keeping open office hours can help increase the flow of healthy communication and increase confidence in the individuals on the team.

The concept behind open office hours in a physical office is fairly simple. You have a set period of time during which employees know you will be in your office and they can walk in and chat with you. They are told that they can come in and talk to you about anything, whether it's project related, a personal issue, questions about policy, or anything else. This keeps up a good flow of communication, encourages mentorship, and creates a forum for people to be open and not feel self-conscious about how they may look to their peers.

Many times in larger organizations, employees who are looking for guidance feel that some managers are unapproachable. The persona of the busy executive oftentimes makes them seem standoffish or distracted by important matters. People who have more reserved personalities may avoid trying to turn to those executives for direction. This would lead to employees having a heightened feeling of insecurity about their jobs and less confidence in their day-to-day activities. The isolation that comes in the remote work environment may intensify these feelings.

By offering open office hours to the organization, management will encourage more natural conversations and opportunities for one-on-one communication. In larger organizations that have a more complex hierarchy, the open office hours will help individuals feel less compartmentalized. Management will see that this opens up the ability to facilitate an exchange of information with people they may not interact with on a regular basis. When that happens, the entire organization becomes stronger and more cohesive.

Just stating your office hours once is not enough for the team to get the full benefit of the practice. Keep your open office hours conspicuously posted where they can be seen by the remote team as a reminder that you are available. Post them in places such as on a group calendar or on an online project board. Set the hours so that people in all time zones are able to take advantage of the policy. Be sure that everyone is aware that *anyone* can come speak with you about *anything* during your open office hours. This creates a feeling of inclusiveness and will alleviate feelings of isolation.

BEST PRACTICES

Leaders should set office hours that work with all team members' schedules. Be accessible and be sure that people feel comfortable coming in and discussing anything.

Make Decisions as a Team

A big part of making people feel as if they are a valued part of the team is allowing them to take part in decision making. When it comes to remote work practices that affect the individuals on the team, empower them by having the team make the decisions, such as choosing the open office hours that work best for everyone or policies regarding flexible hours. Facilitate finding what works best for everyone when possible. This helps makes a bad situation more tolerable and helps employees regain their sense of control in the crisis. These group decisions will both increase transparency and add to the feeling of inclusiveness.

Before you start such a process, be sure to set the ground rules. Will decisions be made by majority rule or must decisions be unanimous? Will there be a person who acts as an impartial tiebreaker? Log team decisions in a shared document and add to it any ideas that were considered but voted down along with the reasons why. This will help serve as a gentle and time-saving reminder should the topic come up again.

Bring Back the Break Room

There is no doubt that social interaction helps to strengthen the bonds within any team. Play creates emotional connections and helps people recognize similarities within each other. With the inability to run into each other in the hallways or in the break room, remote teams miss out on this vital part of the traditional office. It is up to management to help create opportunities to have the social interactions healthy teams need on an impromptu basis or as planned events.

Build Non-Business Activities into Business Hours

In the physical office, teams find reasons to create non-business activities where they would get together. These would happen both inside and out of the office. Birthdays, baby showers, and other celebrations are cultural norms in the workplace. A forced remote work situation does not need to negate these typical social events. Management needs to make a point to bring these occurrences back into the virtual workplace. I spoke with Leslie Murphy of Raybourn Group International about how she helped to strengthen her teams and company culture when they were forced to go remote:

We still do the Monday morning coffees and the Thursday afternoon happy hours. We have gotten a little bit more intentional with those. We will keep doing something because we're really able to engage all of our staff, and we're much more intentional about engaging. Like in celebrations. We will definitely continue to do celebrations, because I think it really helps build our culture, and helps those people feel connected. We did a virtual baby shower for one of our folks and literally everybody in our offices were on their screen. And I would run over to get the present and then run back to my desk with the gift. It just lets people connect in such a different way. And we've noticed with people who are more introverted and probably wouldn't speak out in that kind of other situation (in-person situation) are much more engaged.

Virtual happy hours became commonplace during the COVID-19 pandemic. It was a way for people to schedule having a drink together, unwind, and just shoot the breeze as they normally would pre-shutdowns. Just like a regular happy hour, virtual ones let co-workers decompress together and just talk. It is those simple times that can create the strongest bonds between people.

One executive that I interviewed told me how he arranged for his team to meet up in a public park that was near their office during their state shutdowns. Due to COVID-19 distancing guidelines, they were not able to have any physical interactions, but they were all present with each other. That went a very long way for team morale and reminded them of how much they cared for each other. With many small businesses, the employees are a family. And families miss each other when they cannot be together as they usually would.

Even with being physically separated out of the office, there are ways to help maintain a sense of normalcy to office social interactions. They help bring people together and create opportunity for playfulness and joy. In any crisis, those are two commodities that can sustain teamwork and decreases stress.

Non-Work Chat Channels

Another very popular choice that teams encouraged during the forced WFH experience due to COVID-19 was dedicated social channels on their collaborative work platforms. Social Slack channels, for example, were a huge hit with remote teams. They helped people feel connected and created a space that replaced the non-work-related communication that they had in the workplace.

Extroverts and those who are more socially active in general find the isolation from a forced WFH scenario a bit more challenging than introverts. They need and sometimes crave more human interaction. This is another factor in the human equation that managers and team leaders need to take into account when it comes to the health of the team. In times like these, people who are more extroverted will look for reasons to interact sociality. It helps keep them going. Setting them up for continued success in the remote environment by giving them a medium for that interaction can be a simple act that creates a great amount of good.

Your People Are Your Best Assets

The team is what keeps your company going, but the team is made up of individuals. To keep the team working well and in synch as a cohesive unit, you must focus on care for the individuals. Your people are your strength. Protect them by helping them know when to log off and take a breath. Many times, people will overwork themselves by trying to be a better team player and a stronger asset to their team. Although well intentioned, they can end up hurting themselves by getting burnt out. That ultimately has a negative impact on the rest of the team as well.

Team leaders should be accessible and engage their people to try to identify signs of stress that may lead to larger issues. In the following interview, CEO Wendy O'Donovan Phillips discusses how she discovered that it was necessary to be more active in looking for clues that employees may be struggling in the forced WFH scenario. Taking preventive steps is key to your strategy. This includes reducing unnecessary communications to allow for more focused work as well as reducing meetings. Limiting meetings does not mean that you will get poorer communications; instead, it involves making those meetings more effective.

WENDY O'DONOVAN PHILLIPS, CEO BIG BUZZ

Company Profile

- Location: Denver, Colorado
- Employees: 6
- Primary Line of Business: Big Buzz solves complex marketing problems for the people in senior living, dentistry, and healthcare.
- Primary Audience: Healthcare businesses

About Us

Big Buzz is a healthcare marketing agency delivering focused marketing efforts for executives and teams nationwide. Since 2007, Big Buzz has led clients through the various stages of our disciplined process, depending upon client needs: research, strategy, creation, implementation, and optimization. With this focused process, the Big Buzz team precisely diagnoses marketing challenges, offers custom prescriptions for improvement, and helps client teams realize measurable results.

What are your key in-office processes? Does your team use any unique or key processes in your day-to-day work (e.g., brainstorming meetings, sales calls, group tasks)?

As a marketing firm, collaboration is one of our key processes. Certainly, we can use technologies like Slack and Zoom to help with the collaborative process, but nothing replicates being in the office together where we regularly ran ideas past each other, got ad hoc feedback on creative processes, and acted spontaneously on the need for strategy sessions.

What tech services/software did you use to go remote? What apps, services, or technology did you use to bridge the gap from in-person to remote in order to keep the workflow alive with your team?

More important than what technology we use now to replicate that level of collaboration is how we use the technology. In the early work-from-home days, we used Zoom only in the obvious sense: for scheduled meetings. Now we use it more creatively. For example, we have a standing Zoom happy hour every Thursday afternoon to boost the sense of united culture and camaraderie. We have Zoom teamwork days, where a member of the leadership team logs into a Zoom meeting for a 4-hour period and other team members can join just to work alongside someone else. We all come together for a 15-minute morning huddle on Zoom, where each team member can share personal good news, professional good news, what they are working on, and where they need support. All of these help us more readily put our people before projects and see where we can connect and collaborate.

How quickly did you transition to fully remote? What was the trigger that day you went remote so suddenly?

We went fully remote on March 13, and were well set up to be successful with it right from the start, as we were already using Slack and Zoom. We have just had to deepen that success by thinking outside the box on how to replicate the in-office interactions. It's these interactions—conversations between tasks and meetings—that build trust among the team and bolster us all to take the best possible care of our clients.

What was the hardest part of going remote for you and your team?

The most difficult part has been maintaining our collaborative culture.

What was the easiest part of going remote for you and your team?

The easiest part was use of technology like Slack and Zoom. Zoom hackers were an unexpected surprise. I've been in Zoom meetings where hackers yell racial slurs, go on cussing tirades—I even watched a guy pretend to shoot himself on camera. Very unpleasant stuff. Glad to see Zoom is cracking down on it.

What part of your management style needed to be tweaked with a remote team? What are some things, if any, that you discovered you had to change with regard to how you interact with your team?

I am a driving, focused manager: do the task, solve the problem, finish the plan. In this day and age, I now better understand the softer side of leadership. It's important I listen to my team, be empathetic to the challenges they are facing during this difficult time, say and do the right things to unite them, and build a culture of trust. Those soft skills are actually more important than the hard skills. When people trust the organization, leadership, and each other, they produce better work.

How did this affect your company culture?

It's just been more challenging to maintain that collaboration and build in time to work on those soft management skills. In the office, I can see and sense when someone is not well or struggling. Remotely, it's tougher to identify, so we have had to rethink ways to meaningfully connect.

How did you combat the feeling of isolation in your company? Can you describe briefly ways you came up with to combat isolation? What did you add to help keep people feeling connected?

The three remote meetings are helping: weekly happy hour, occasional team-work days, and daily 15-minute check-ins. We also love www.window-swap.com. Check it out, and be sure to turn up the volume. You won't feel alone anymore.

Briefly summarize what you may do a bit differently (if anything) if you had to suddenly switch your organization to working remotely or if you could do it all over again?

I would have added the three remote meetings sooner. I would have worked more readily on honing the soft skills of management in me and my leadership team.

What words of wisdom and advice could you give to someone else who finds themself having to suddenly make their operations go remote? Any pitfalls that people should look out for?

I'm not big on giving advice. (Advice is criticism dressed in a cashmere sweater.) I do hope some of the experience and learnings I shared above might benefit others. Thanks for having me!

Client and Team Meetings: Making the Most of It

We have discussed many ways virtual meetings can be exhausting. Zoom fatigue is a real phenomenon, and too many remote meetings can kill productivity. Some meetings need to happen to get people on the same page, but arguably, many times that can be done with other, more asymmetric methods such as a voicemail, email, or a note pinned on a virtual project board. But there are some meetings that are still necessary, and meetings are not inherently a bad thing. In fact, a good meeting can be invigorating. There is a human comfort factor involved. A good meeting makes people feel that everyone is on the same page. The official meeting, in many cases, acts as an affirmation of the team's or client's feelings that everything is on track.

As has been observed, virtual meetings lose elements that are only found in face-to-face meetings. It can take several remote communications to get across the same level of understanding you would get in a single in-person meeting. One reason is that in-person meetings and virtual meetings require different skill sets to achieve the same results. Real-world methods don't always translate well in the virtual world.

Being brought up in a multilingual household, I learned early on that many idioms and idiosyncrasies do not translate from one culture to the other. I found this to be true when working with developers in other

165

countries. I adjust my speech and written communications to take out phrases that might be common to me but may be confusing to others. I am more deliberate in the way I use language as a tool. Think about the difference between a machete and a scalpel. Both instruments can be very useful for their own purpose, but you would not use them interchangeably. The same concept applies to real-world skill sets and virtual-world skill sets when it comes to meetings and getting your point across. One requires much more precision in how you use your tools and requires closer observation if you are to have the same level of success. In this chapter we discuss how to make the most of your virtual meetings with your team, clients, and others outside of your organizations.

A Quick Case for Virtual Meetings

Prior to the COVID-19 pandemic, videoconferencing was used but not to the same degree that businesses had to when most employees started working remotely. Many small businesses did not initially use video to conduct business but quickly learned to do so. When your company switches to a remote work situation, it may be due to a temporary need. But I would encourage you to make video meetings a default setting for your company.

As many businesses have noted, there have been some unexpected and helpful operational advancements from the abrupt shift to working remotely. Some companies such as Raybourn Group International already had remote workers on their team, and those employees were likely to never be in the office. However, that meant that the team would never have an opportunity to meet each other in person, which may have seemed unusual to some. Prior to the 2020 shutdowns, video was not the norm for team meetings that included remote members. Here is some insight from Leslie Murphy, CEO of the Raybourn Group:

> Sometimes if we had a team meeting, people would meet in the conference room and then they would audio in the person who was remote. We're not going to do that anymore, because we found that when we're all on the same level playing field, people engage much better.

In other words, instead of having a remote worker phone in, they will be using video as part of their team meetings from now on. They have also decided to leverage video meetings for other operations of the company including new hires, as Leslie indicates:

We utilized Zoom for all interviews and advertised we were doing that to help people feel comfortable about the process. We will always utilize Zoom or Teams in the future, at least for first interviews. The benefit was that we were able to engage more of our remote and non-Indianapolis-based staff, which was very beneficial to the process.

Clearly, for hybrid teams that work in office and remotely, sticking to a video format keeps everyone feeling engaged and on the same level. But it has to be done the right way. Having everyone who is in the office sit around the conference table while the remote workers log in from their various locations isn't optimal. That only amplifies the feeling of being disconnected. It projects the idea of two teams, those on the inside and those on the outside looking in. Instead, everyone should be in their respective spots joining in from their workstations.

I Stopped in 2014

Now that video meetings are part of everyday culture, you should also reconsider your need for face-to-face client meetings. Video meetings are now the norm for every level of business, so virtual client meetings are no longer unusual but are widely accepted.

In 2014, I made my last business trip to meet with clients. I found myself being much more stressed when I was traveling for business than when I was working from home or my office. When I am on the road, I always felt as if there was work that I was missing or couldn't get done while I was in the airport, on the plane, or being taxied to the hotels. Some of that was built up in my mind, and those feelings were exaggerated by the fact that I would much rather be nearer to my family and pets and the comforts of home. I am certainly not the only person who feels this way. There are countless articles on the subject of reducing stress while on business trips.

Then, of course, there is the expense of business travel. Figure 7.1 shows data from a Verizon Business report on the financial and stress-related impacts of business travel for in-person meetings. This data shows that companies should consider whether traveling to meet with clients in person is that much more advantageous and worth the expense. There are those who still argue that nothing beats the face-to-face meeting.

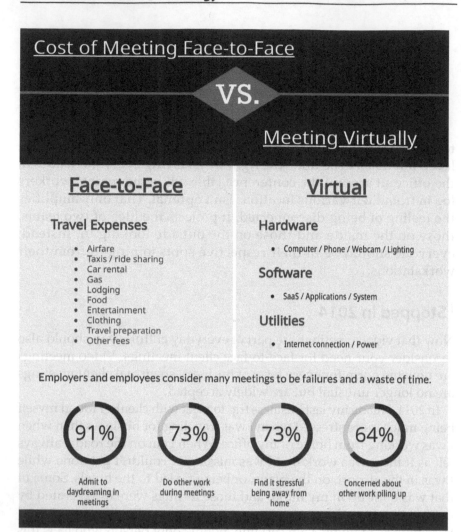

Cost of Meeting Face-to-Face

VS.

Meeting Virtually

Face-to-Face

Travel Expenses

- Airfare
- Taxis / ride sharing
- Car rental
- Gas
- Lodging
- Food
- Entertainment
- Clothing
- Travel preparation
- Other fees

Virtual

Hardware

- Computer / Phone / Webcam / Lighting

Software

- SaaS / Applications / System

Utilities

- Internet connection / Power

Employers and employees consider many meetings to be failures and a waste of time.

91% Admit to daydreaming in meetings

73% Do other work during meetings

73% Find it stressful being away from home

64% Concerned about other work piling up

Figure 7.1: Cost of meeting face-to-face

The Personal Relationship

It is ironic that, when I chose to stop traveling to work, a friend of mine wrote a blog on the merits of in-person business meetings that got a good deal of media attention. In-person meetings with clients are still a good idea, but they should not be your default setting. One major benefit of in-person meetings with clients is building personal relationships. I fully agree that much of business is built on relationships, but we can build extraordinary relationships with people even if we never meet face-to-face. The proof of that came during the pandemic.

My company works very closely with our clients. We are a B2B company that helps with our clients' digital marketing and customer engagement. We get to know not only our customers' brands very well, but the individuals behind the brand. During the shutdown, we received calls from very emotional clients. They opened up to us about what the shutdown was doing to their business, including their fears that they would need to lay people off or even shutter their business for good. They were afraid not only for their employees' future, but the future of their own families and themselves. Some even openly wept with us on the phone.

We are close with our customers—close enough that they feel comfortable being open and vulnerable with us. They feel that they know us as people well enough to be able to tell us that they are afraid of what is happening to them as individuals as well as business owners. And yet, we have never met them in person.

Virtual client meetings can be just as effective as in-person client meetings. You can still use remote technologies to build great client relationships and relationships with co-workers. One of the keys to this is to always be authentic. Be true to your values and your company mission, and don't be afraid to show your human side. And remember, virtual meetings require separate skill sets from those needed for in-person presentations.

Before You Start

Hopefully you would never simply go on stage in front of a crowded room to give a presentation without being fully prepared for it. Even if you are fluent in your subject matter, you wouldn't want to just wing it. You would want to have your facts and figures ready, and you would want to know what aspects of the topic you will cover and how they are related. You would prepare any images that emphasize your points and you would want to be sure you were up-to-date on your information so that you can respond intelligently should anyone from the audience ask any questions. These line items on the preparation task list cover the presentation information, but there are other important tasks you want to check off prior to walking onto that stage.

Consider for a moment the place where you are speaking. The stage has to be set, the lighting needs to be in place so that people can see you well, and the audio needs to be checked to be sure that the microphone works and the speakers emit a clean sound so that you can be heard as well as understood. Your images will need to be projected with equipment or printed so that you can share them with the audience. You need to prepare yourself as well. You should be showered and well-groomed. You will want to choose appropriate professional clothing that projects your expertise and authority on your subject. Then you also need to prepare yourself mentally. If you just had an argument with a spouse or loved one before the meeting, you need to push those issues and emotions out of your head for a while. You need to center your mind and find clarity so that you won't be distracted by outside influences.

These are some of the steps you would need to take if you were to have a successful on-stage presentation, and you should keep them in mind to help make your video meetings successful as well. Virtual meetings have become so commonplace that many times people do not take the time to set the virtual stage or prepare the other aspects of the meeting that one would have done in a more formal setting. Here, we discuss some of the basics that you should consider in order to have a highly successful virtual meeting where people feel engaged and you maintain their attention on the topic.

Keep Them Engaged by Design

One important topic that many business news websites discussed during the shutdown was how easily distracted people were on video meetings. This speaks to the high rate of fatigue people were experiencing on video meetings. Many speculated that it was related to the pandemic. Others considered whether millennials and Gen Z had shorter attention spans than Gen X or baby boomers. Still others attributed it to the large number of video meetings overall. But being distracted during meetings is nothing new, and it's certainly not limited to video meetings.

Figure 7.2 and Figure 7.3 illustrate distractibility during virtual meetings. The first is from a 2014 *Harvard Business Review* article titled "What People Are Really Doing When They're on a Conference Call." The second graph is taken from a June 2020 report from online career and job website Zippia entitled "Most People Are Distracted During Virtual Meetings." The first of these articles focuses on phone calls, and the second focuses on video meetings. These sets of data were strikingly similar despite the six years between articles and the different technologies being utilized. Video calls were not commonly used in 2014; most companies used speaker phones in conference rooms for their conference calls.

What People Do While On A Conference Call

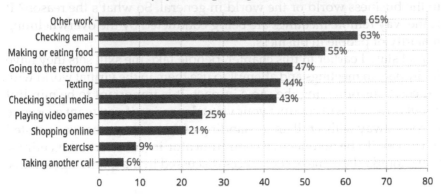

Figure 7.2: What people do while on a conference call

What People Do While In A Virtual Meeting

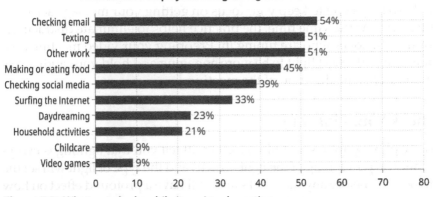

Figure 7.3: What people do while in a virtual meeting

In 2014, there was no great upheaval in the business world as there was in 2020, yet people were still just as distracted during remote meetings. It is unlikely that people were widely reported to be so distracted in meetings during the pandemic simply because of the large number of video meetings. In 2014, there were generally far fewer meetings than in 2020 because many additional meetings had to take place to get everyone on the same page while teams were widely distributed during work-from-home periods.

So, the problem of people not being engaged in their remote meetings is not the type of technology. There may have been fewer meetings in 2014, but there was still a high level of distracted people who were not fully engaged in those remote meetings. The issue does not seem to simply be the number of meetings taking place, and it's not a sudden upheaval in the business world or the world in general. So what's the reason? It is the way remote meetings are being conducted by the overwhelming majority of people in business.

As I stated earlier in this chapter, people take the skill sets they have for in-person meetings and attempt to use them in the virtual world with expectations of having the same impact. It is important to be mindful about how you will conduct your virtual meeting if you want everyone to come away with a full understanding of the message that you wanted to convey. To be an effective communicator in your virtual meetings, you should follow the best practices discussed in the following sections.

Prepping Your Space

Preparing your physical space for your meetings needs to be a fundamental practice for the virtual office, as it is in the real world. Taking the time to do so helps you to keep your focus on getting your message across to attendees. At the same time, the practice helps to eliminate distractions for the attendees who are tuning in. Prepping your video meeting area puts you in the best light to be more engaging and hold your audience's attention for the duration of the meeting.

Check Your Lighting

Many photographers and filmmakers will tell you that lighting is every-thing. The positioning of the light sources and the type of light will set the mood even before anyone speaks and will have a profound effect on how people perceive the subject matter as well as the message that they take away.

I spoke with professional lighting designer Laura Glover, who has spent over 30 years lighting the stage for professional theatrical productions and events. We discussed the impact that lighting has on the speaker as well as the audience. I asked her if she could share some knowledge and best practices when it comes to lighting for video meetings:

> *Good lighting for video calls involves seeing the participant's face clearly so that we can see them talk. Just like seeing an actor on stage well enough so that you may hear them correctly, we need to see a participant's face in order to hear them well enough. A bit of dimensionality, the partici-pant popping forward from the background, is also helpful.*

Bad lighting is full of shadows, dim faces, too-bright backgrounds, and distractions from a too-busy background. It's all about attention on the speaker. If you cannot see the speaker clearly, you will have trouble hearing what they say. If the background competes with the speaker's face, then your mind may wander to thoughts that take you out of the conversation.

The ring light is great for lighting a speaker's face. They come in all sorts of sizes, with cell phone tripods and other accessories. You can pick one up for $20 or for ten times that price. Mount it behind your computer, turn it on and—voilà!—the speaker is lit well with a diffused light that softens the face. Also pay attention to color. Some ring lights come in different color temperatures. I find having one that is in the midrange around 3100 to 4500 Kelvin is good and natural. Warm white will make the face muddy, and the cool white—well, you may end up looking like a vampire! Play with the color by looking at the camera view of you before you attend a meeting. You and I, because our skin tones are different, are going to look very dissimilar under the same temperature light.

In terms of popping out of the background, you could actually light the wall behind the speaker with a soft light (a clip light or two on the floor is a cheap trick). It is important that it is not intrusive. Wall art is fine behind the speaker, as long as it is not too busy. I've been on videoconferencing calls where one of the attendants likes putting up another location behind them (i.e., the Golden Gate Bridge or space), and I find that incredibly distracting.

Do not have anything reflective behind the speaker! Whatever light is lighting the face will reflect in the glass and be equally distracting.

Do not sit in front of a window. The speaker will be competing against the light from outside.

A blank wall is equally distracting to me. Avoid that. People like to know a little bit about who they are speaking with. Give them an idea but not all the knowledge!

That is really about it. Pretty simple: Good face light to see and hear the speaker. Pop from the background. Do not distract.

Design Your Background

During the height of the COVID-19 pandemic, Twitter, and specifically the Twitter account for Room Rater, got international media attention for judging the rooms where celebrities took their video calls. It was a popular way to gain insights into the lives of famous individuals. It speaks to the fact that people do in fact take notice of the room you are in and what's around you during your virtual meetings. One of the reasons behind this is not simply curiosity, but it's because the environment where you conduct your meeting gives people insight about *you*.

It is true that a home is a reflection of those who live there. If you were to have a client meeting in your place of business, both the meeting room and the overall environment of your office would expose your clients to your company's natural habitat. It would reflect your brand and your company culture. If you are hosting the meeting from your home, your background should still reflect some of the elements that your company may bring to the table. Consider whether it should be orderly and efficient or creative and eclectic. These are all elements that attendees will subconsciously associate with you and your company when they are invited into your own environment for your video meeting. You can choose what your room says about you and your brand by designing it with purposeful intent.

You can, of course, use a digital background, but those can give off the appearance that you did not invest a good deal of time into the meeting, and they can be cludgy if the speaker moves around and the digital image blurs or gets spotty. If you can make an interesting background in the room where you are taking your meeting, do that instead of using a virtual background. It's worth the investment of your time to make the effort.

A few decades ago, it was not uncommon to invite a client or your employer over to your home for dinner. In fact, it was a common enough event that dinner with the boss is considered a TV trope. There was good, practical reasoning for this now outdated dinner practice. Inviting people into your home is an opportunity for you to showcase your interests outside of work while remaining in your professional mode. This is still true today even if it is for a virtual meeting. It's an opportunity for people to get to know another side of you. Use your room and background to take advantage of this opportunity, but be sure to keep it simple. It should look natural and not staged. There is no need to add interesting items behind you so that the background looks cluttered or distracting. You want the focus to be on you and your message. But you need to be mindful that people will still be taking note of what is behind you and around you.

For internal meetings, tools such as Together Mode from Microsoft Teams can help people from your own team feel more connected. The feature allows for your attendees to see each other occupying the same virtual space, such as a coffee shop or auditorium. This puts everyone on the team on a more even keel, and it takes pressure away from them to have an interesting background in their own environment.

Declutter

When you are speaking, you want your audience focused on you and your message. Remove any visible clutter from your workspace. You want your stage to be free from things that may pull focus away from you or your message. This also lends to your credibility. A disorganized setting with piles of papers or other items strewn about does not inspire confidence.

At the same time, you want to make sure your environment is not distracting to *you*. Remove clutter from the other side of the camera as well. Seeing a stack of papers or other items that have nothing to do with your meeting may draw your attention away. You want to remain focused on the task at hand for as long as possible. Glances at piles of work that still need to be done will be a distraction. Once you get this done, you have the added benefit of a clean and organized work area to use after your meeting.

Prepare Your Devices

Getting your devices ready for a meeting will ensure that you can be heard clearly and be seen without the menace of lag, stutter, and glitches. Take a moment to update your device drivers if you have recurring lag issues. Running updates on your devices regularly will help further eliminate meeting snafus.

Remove Online Distractions

You don't want your computer or phone to interrupt your meeting and cause a hiccup in your presentations. Close all tabs that you won't be using for the meeting to remove distractions and free up resources on your computer. You want good processing speeds, and you want to eliminate any potential lags in your video.

Turn off notifications on your computer. I've been in video meetings where the presenter was sharing their screen when suddenly they had a notification of a new email that popped up on their screen with an embarrassing subject line. If you also use a tablet, turn off your notifications on that as well as your phone. All of your devices should be in a mode where you won't get any alerts during your meeting. This is a great time to use one of the focus apps that we discussed in the previous chapter.

Kill the Bandwidth Eaters

You want to reduce the possibility of any video or audio lag for your meeting. Not everyone has great bandwidth at home. Most people do not have the amount of bandwidth at home they have at the office. If the Internet connection is not great where you are, make sure you turn off anything that may eat away at the connection and slow it down. Disconnect your phone or other devices from your Wi-Fi and turn off any apps on your computer that you won't be using. Ask roommates or kids at home not to stream video or play games online during the time that you will be on your meeting.

Test Your Audio

It is very important to make sure your audio—both the microphone and speakers—is working properly. I have made it a habit to use the testing function in the video app to make sure the microphone and speakers are in working order. You want to be able to hear others in the meeting, and you want to be both heard and understood. If your mic is blocked by papers or something else, you may come off as sounding muffled.

This is also a good reminder to make sure the settings in your email for the meeting state the correct way to call in for the meeting. Some app integrations automatically will set up a phone number to call into or a link to click to join the meeting. I've had some meeting invites that listed both a Zoom and a Google Meet link because one of the integrations added its own call-in line. Such duplications can cause confusion and can result in some attendees showing up late or missing the meeting entirely.

Find the mute button and other audio settings for the meeting attendees before you start. If you need them, you do not want to look as if you are fumbling to find them. That's a quick way to become suddenly self-conscious in front of your audience and lose your focus for the meeting.

Eliminate Background Noise

If you are working remotely, you may have little control over your environment. Roommates, children, pets, and even neighbors can cause unplanned interruptions that your mic could pick up and pass on to your meeting audience. Eliminate the need to apologize for those interruptions by being proactive and using a background noise filter app.

Krisp

Krisp is a noise canceling app that works across platforms, so it's suitable for use on any device. You can use it to reduce background noises on a video meeting conference call or even for recordings that you may use for asynchronous messages.

SoliCall Pro

SoliCall Pro reduces background noise on VoIP calls. It also eliminates the echoes that you sometimes get on VoIP calls. This service works on both sides of the call, even if you are the only one using the app. This is a great app for conference calls or any business calls over a VoIP system.

Change Audio Settings

There are changes that you can make within your video meeting service to help suppress background noise interruptions. For example, in Microsoft Teams there is a setting for noise suppression, and in Zoom there is a tab to suppress background noise in the audio setting section. Get familiar with these more advanced settings. It's best to be able to take full advantage of the features of the services that you are using.

Adjust Your Camera

You want to make sure you have both your camera angle and settings adjusted properly. Think of how a news anchor appears on the screen during a broadcast. You don't want your face to take up the majority of the screen as you may see on video calls from a smartphone. News anchors have the shot positioned in a way that you can see their head and shoulders; you should adjust your camera in the same manner. Set your camera at eye level and don't be too close.

Don't limit yourself to only using the built-in camera on your laptop. Invest in your meetings and your company. You want to present yourself in the best way, and that means using a good camera. Even on expensive laptops, the built-in cameras are not nearly as good as the ones in your phone. Video quality from laptop cameras is not going to be optimal, but there is no need to feel as if you are stuck with the tech that came with the computer. Many savvy businesspeople purchase an external web cam so that they can showcase their business in the best light.

Get something with a minimum of 1080p (meaning a resolution of 1920×1080 pixels). Even though a resolution of 720p is considered high definition (HD), the video will not be nearly as clear as 1080p, which is considered full HD (FHD). Every digital image is made of a mass of tiny squares called pixels. Each one helps to build the larger image. The more pixels an image holds, the more information or detail the image will have. So, an image taken with a 1080p camera will have better clarity and seem crisper to the eye than an image taken by a 720p camera because the 1080p camera can take images that hold much more detail.

There are web cams that go beyond 1080p to 4K, or Ultra HD. In most cases 4K web cams would be overkill. But do shop around, and take into consideration that some of these web cams will also come with built-in microphones. There are some great features on these web cams, such as noise cancelation, auto-focus, tilt, pan, and zoom.

There are also features on the video services that you can use to help you look your best. News anchors and other on-camera personalities use makeup to help smooth their skin tone and brighten their appearances, and video software is following their lead. For example, Zoom has a feature called "touch up my appearance," and it does exactly that. It will make your skin seem smoother and will remove some blemishes and wrinkles to help you look and feel your best on camera. That's a great feature to use because the better you feel about your appearance on camera, the stronger your confidence will be. Self-confidence is something that is noticeable and is radiated out to your audience. So, take a good look and see which features on both your camera and the service you can use to leverage them to your advantage. Some simple adjustments can make your meetings even more engaging and help improve your presentation.

Prepare Yourself

Having the best working equipment and a pristine meeting space cannot help if you, yourself are poorly prepared for a meeting. In fact, the great equipment and nice space will only amplify the fact that you are underprepared. Besides rehearsing what you will discuss in the meeting, there are some additional steps that you can take to get yourself ready for the camera.

Dress for the Meeting

Remember to dress for the part. This speaks of your credibility as well as your commitment to what you are doing. If you wouldn't wear it to the office, don't wear it for your virtual meeting.

Avoid wearing stripes or patterns. They can be distracting and don't always show up on camera as well as they do in face-to-face encounters. Instead, use solid colors to make you seem more authentic, and choose those colors wisely. Color psychology is used in marketing without most consumers being aware of it. You may notice that fast food chains all seem to use yellows, oranges, and reds. They do this on purpose because yellow is comforting and cheery, and red makes you pay attention, invokes impulsiveness, and triggers the appetite.

Bold colors will give you an air of confidence, but you do not want them to be too bright. Wearing overly bright colors may trigger your camera's auto-focus feature and cause it to attempt to keep refocusing during your meeting. A bright white shirt may look great in person, but on video, a more muted off-white would be better. Avoid wearing black or dark colors. Dark can set the tone to be gloomy. You want your meeting to be engaging and inviting.

Posture

At the end of this chapter is my interview with Mark Edgar Stephens, who is an expert in body language and communication strategies for executive leadership and their teams. We will be going into more detail in that interview about the importance of your body language, and non-verbal cues, but for right now we will focus on posture.

Before you speak a single word, your posture will be speaking for you. It will broadcast your mood, your confidence, and your readiness for the meeting. Be aware of how you sit and adjust your body to optimize your appearance. You should sit up straight, head facing forward, with your ears, shoulders, and hips aligned. To help you keep this good posture, sit with your feet in front of you with your knees bent at 90 degrees.

Know Your Attendees

Regardless of whether you are meeting with 8, 80, or 800 people, you should know your audience. The meeting is all about them, so you will need to know who they are in order to get your message across in a way that hits home.

If you are presenting to a client remotely, then get a list of your attendees and look them up on LinkedIn or do a web search on them. You can find out where else they may have worked and in what other positions, as well as some of their professional interests. This audience-centered approach will assist you in tailoring your message in a way that will pique their interests and keep them engaged so that you can more effectively deliver your message.

Get Some Clarity: Quiet Your Mind

If you have any calming or breathing exercises that you do in order to help de-stress and relax, do those prior to your meeting. Many professionals find that doing such exercises helps them not only relax but also retain their focus during meetings.

I also try to limit my caffeine intake for the day prior to any meetings where I am the presenter. I tend to be a bit of a fast talker, and when I am passionate about my topic, I tend to get excited and speak quickly. Having high amounts of caffeine in my system exacerbates that issue.

Prior to the meeting, have your mind focused on what you want to accomplish in the meeting and the outcome that you are working toward. Visualizing your intended outcome before the meeting is a technique that many swear by for good outcomes. Think of it as a way of keeping your eye on the prize. Many times, we get so caught up in the details that we forget about why we are preparing so carefully. Keeping your mind's eye on target will help you maintain your composure and keep your thoughts sharply focused on your goal of the meeting.

Etiquette

It is essential to follow proper etiquette for meetings, if for no other reason than to keep things organized and productive. I learned this well in my time serving on committees for nonprofit organizations. Nonprofits tend to use Robert's Rules of Order for board meetings and some committee meetings. It helps them to do the business of the board efficiently by setting up ground rules for how many board members must be in attendance for the meeting to do business or take votes. It also helps define who gets to speak and when they get to speak, along with what items will be discussed in the meeting agenda.

Understanding proper etiquette for video meetings in business also helps establish an order to the meeting and keeps it productive. There are general rules and then some that are geared toward being an attendee or being a presenter.

Check Your Tech

Of course, checking your technology before joining is important. Make sure your speakers and mic are working so you are not a distraction or wasting time trying to get them on once the meeting begins. If you are using a wireless headset with built-in mic, be sure that it is already wirelessly connected to your computer, and do the sound test for the speakers and microphone. Making sure your tech is working before the meeting begins is a good segue into the next point of etiquette, which is get there early.

Arrive Early

Regardless of whether the meeting is virtual or in person, it's a good idea to get there early. This gives you the opportunity to decompress prior to the start, whereas getting there just in time gives you no time to settle in. Arriving a bit early also allows for some niceties and connections to form between the other attendees.

Make Eye Contact

Another good point that does translate well from the real world to the virtual world is to look people in the eye when speaking to them. If you are looking at the screen, it won't seem as if you are looking into the eyes of the person you are speaking to on the other side of the screen. In order to give that appearance, you must look into the camera lens. This is a good reason to position the lens ahead of time to help optimize effective communication opportunities in the meeting.

Cut the Mic

Mute your microphone when you are not speaking. This helps cut any distractions from background noises that could interrupt or draw attention away from the speaker.

Use Their Names

Call people by their names when you are speaking to them in a group meeting. Not only does using a person's name command their attention, it's a way of recognizing them. It is more engaging. It shows that you see them as an individual, not just as one of the people from the team.

Pay Attention

It's unfortunate that this has to be said, but it happens far too often in video meetings. Not paying attention to what's going on in your meeting is rude regardless if you are an attendee or one of the speakers. It's disrespectful to the individuals who are speaking, as well as to the other people in the meeting, and it makes you look bad.

Put down your phone and keep your screen on the meeting instead of going to a page that has no relevance to the meeting. If you use a focus app or turn off notifications and close your other browser tabs as discussed earlier, it should not be difficult to avoid distractions.

When You Are the Presenter

If you are presenting, send out the meeting agenda ahead of time. This helps make sure nobody is going into the meeting cold or does not understand the scope or depth of the topic. Be sure to stick to this agenda and the schedule that you have set. Start on time and end on time.

If there will be action items or decisions to be made during the meeting, spell those out ahead of time in your agenda. This creates the opportunity to ensure that everyone is on the same page before stepping into the meeting.

Start off the meeting with gratitude. Thank people for being there and coming to your meeting. Everyone values their own time, so if they are taking the time out to meet with you, it is common courtesy to thank them for it. If appropriate, then make introductions.

Add in pauses if there is a good deal of information to go over. Taking a moment to ask if anyone has any questions prior to moving on the next topic helps ensure better understanding. To that end, you should record the meeting if that feature is available in your video service. If not, engage someone in the meeting to take notes, then make those notes available, if appropriate.

If you want people to be able to chat during the meeting or even go off to side chats or breakout rooms, state that plainly at the beginning. Then make sure those settings are enabled.

Before everyone breaks and the meeting ends, be sure to review any action items that were discussed. State who is assigned what tasks and who is accountable to ensure those items are acted upon by a specified date. This helps ensure clear communications and expectations.

When You Are an Attendee

Be sure to have something to eat prior to your meeting if it will take place close to when you would normally have a meal. You do not want to be hungry and distracted during the meeting, and you certainly do not want to be eating during the meeting unless everyone is sharing a meal together.

Do your best to keep your eyes on your web camera lens and avoid looking at yourself on screen. It helps others feel that you are engaged in the meeting, and it will help you to not get caught up by distractions of your own appearance.

Do your best not to interrupt if you have a question. If you feel you may forget, write it down. Then read the room, and if there is an appropriate pause, ask the question. Many video services have a feature where you can virtually raise your hand to notify the meeting planners or presenters that you have a question. Use that if it's available, unless the presenter states ahead of time that they want everyone to hold their questions until the end.

If for some reason you do not feel comfortable being on video for the meeting, reach out to the appropriate person well ahead of time to see if it's acceptable for you to join via audio only. Otherwise, do your best to be on camera.

Make Use of Whiteboards

According to the Soloman-Felder model of learning styles, roughly 65 percent of the population are visual learners. That means that they understand and remember things better if they see the information. For visual learners, seeing the ideas behind the words helps them to make a better connection to the message that you are trying to get across. You may have seen how well this works in a brainstorming session where people are getting up and going to the whiteboard to show the team what they are talking about. Visuals help get the point across faster, and it helps a large portion of people to retain that information. Use those facts to make your video meetings more engaging and more effective with the use of interactive whiteboards.

You can use a digital canvas just as your team would use a whiteboard in person. This tool helps mitigate the hurdles of remote collaborations when it comes to expressing ideas and getting the message across. A digital canvas is an effective collaborative tool. Anyone in the meeting can take over the virtual whiteboard and add on to the ideas just as if they were in the conference room together. A whiteboard can help convey big picture ideas that can then be refined and scrutinized with ease by the group. Videoconferencing giants such as Zoom and Microsoft Teams have built-in whiteboard features. There are also a number of virtual whiteboard providers who have some impressive features to help companies thrive remotely.

Bluescape

Bluescape helps teams collaborate better by making things visually accessible for everyone. In their virtual workspaces, Bluescape allows teams to share not only a whiteboard, but video, images, spreadsheets, and files of all kinds. You can put items to discuss right in front of co-workers just as if you were doing it in person.

The platform works great on laptops, tablets, or phones and even on large interactive touch screen displays. The system will save and store all of the notes and markups from your meetings so there is a record to look back on. It is a great tool for both internal meetups or presentations for clients.

Conceptboard

Conceptboard helps teams to brainstorm, collaborate, and present new ideas with ease. Team members can use sticky notes, line connections, and freehand drawing just as they would in person. The system can make videoconference calls and can screen share right from the app, so it also works well on mobile phones.

Conceptboard is very easy to use, so getting started is fast. It comes with numerous templates that help teams get up and using it right away. It integrates well with Google Drive, Dropbox, Trello, and Microsoft Teams.

Draw.Chat

Draw.Chat is a simple online whiteboard. Its elegant design allows teams to work together quickly and easily. There is no learning curve to use this service. It's highly intuitive and brilliant in its simplicity.

You can get started in seconds. For example, you can drop the URL or link to an image you've found online into the bar in the center of the page and click the button underneath that says "click and draw on picture." Once you click that button, you will be redirected to your new whiteboard with the image and a dashboard of your drawing and note-taking tools. It hosts a number of real-time image editing tools and file storage and file sharing options.

Hoylu

Hoylu's interactive collaboration space allows companies to share ideas in real time and visualize them with ease. Use drag-and-drop customizable templates to quickly build work environments to meet the task at hand. Project planning, brainstorming, project reviews, and workshops are simple to do remotely on Hoylu.

It creates workspaces that are saved and can be retrieved at any time. This means that any workspaces that you have created, or ones that you have been invited to join, can be accessed to get info or to update. It integrates well with Zoom, Cisco Webex, GoTo Meeting, and Microsoft Teams.

Miro

Miro offers an infinite whiteboard that will never run out of space. It's designed for cross-functional teams, so it is robust enough to meet the needs of everyone from marketing to IT teams to use and collaborate with each other. Regardless of whether it is mapping out the customer journey or the steps and tasks needed to launch a new product, this platform is set up to help any size company collaborate remotely and do it well.

It allows users to integrate their documents, screen shots, video, sticky notes, spreadsheets, or any information that may be used for meetings. This allows for there to be a single space that contains everything the team may need to collaborate on a project. This is a feature-rich system that scales well for small teams and enterprise-class companies alike.

Engagement and Communication

The keys to any successful meeting or presentation will be in your engagement and ability to communicate. When you're meeting remotely, those skills must be realigned to meet the needs of a remote audience. You will not be able to easily pick up on nonverbal cues if you do not take the time to learn what to look for in the audience members who are not in the room with you.

By taking the time to consider the steps outlined in the chapter, you will be well set up to keep your remote meetings engaging and effective. Preparing yourself, setting the stage, and investing in the right technology will help you make the most out of your virtual sessions.

However, there is still more to consider when you are communicating remotely in order to maximize the effectiveness of your message. That is why I reached out to executive communications strategist Mark Edgar Stephens for this chapter's interview.

MARK EDGAR STEPHENS

About Mark

In the media, Mark Edgar Stephens has appeared as an expert in his field on *The Oprah Winfrey Show, Dr. Drew, CNN/Headline News, Home & Family, Access Hollywood,* NBC, ABC, FOX, OWN, TLC, Lifetime, Hallmark Channel, MTV, VH1, HGTV, Style Network, Oxygen, and REELZ channels.

He is a frequent keynote speaker for multinational companies and offers workshops and seminars to organizations around the world. From 2016 through 2020, Mark has been honored as one of the "Top 30 Body Language Professionals in the World." He is the author of the book *Who Are You Choosing to Be?*

Mark is a personal and professional development coach and consultant who specializes in body language and communication strategies for executives, leadership teams, and organizations. He has worked extensively with teams ranging from nonprofit organizations to Fortune 500 companies.

As I interview business leaders, I find the same recurring phenomenon that they experienced when they suddenly had their team going remote: It suddenly takes several video meetings and some online chats to get the same message across in what would have previously taken a single in-person meeting. Why is that?

I'll boil this down to four major factors. First among these is attention and focus. In our remote environments, we often have more distractions than in a single, dedicated conference room where everyone is sitting around a table, giving their attention to the same subject, the same speaker, or the same leader. We can see when someone stops listening and starts answering texts or emails on their devices. By being physically present in the room, we can read the body language of fellow team members and know whether they are "with us" or not.

In our home environments, we may have one child fighting with another or a spouse walking about in "less than corporate" attire or be alerted by the smell of a pizza burning in the oven (again!). In the virtual world with larger teams, it becomes extremely challenging and, in some cases impossible, to split our attention between the virtual meeting on our screen and the events happening in our actual physical surroundings. This makes it far too easy to turn off our video monitor or to mute our sound, while we scream for our children to stop writing on the walls. It can be overly tempting to sneak a peek at our computer screens or smart devices to check our social media or to doom scroll.

Second, there is the phenomenon I call "corporate invisibility." Time and again, I've seen large group meetings of six or more people become dominated by one or two people talking, leading, and playing with possibilities, while others, assuming that their input or presence is less important, simply "tune out" from the conversation, hoping to jump on board after the important decisions are made by the leaders or by the more dominant team members. When three people are meeting, it is evident whether the whole group is engaged and participating. The larger a group grows, the easier it becomes to fade out and become invisible. Now add the extra cloaking layer of a larger-scale virtual meeting and we have the perfect recipe for a team-wide spread of corporate invisibility.

Next, there is a factor that so many people never take into account: variations in communication styles. People receive and deliver information in a variety of ways, and what may be satisfactory in regard to understanding for one person may be confusing or even unintelligible to another person. One of the things I do when I work with a leadership team is to identify the nine different communication types on my Communication Styles and Strategy Grid and teach teams how best to flex into and adapt into the communication styles of their leaders, peers, and direct reports, especially when communicating virtually.

Finally, and this may be obvious, there is less body language communicated in a virtual meeting. It is theorized that body language makes up over 50 percent of how we communicate effectively and that the words we use make up less than 10 percent of that overall effectiveness.

Through computer screens, we are usually observing interaction from the neck and shoulders upward, denying us the opportunity to see the full picture being communicated by the rest of the person to whom we are speaking. Now, imagine someone on the other side of the camera with only their face and neck visible, turning toward and away from the camera, sometimes stepping away from or turning off the camera or, as is often the case, the lighting of the person is so bad that their face can barely be seen at all (this happens far too many times).

As a result, the communication intake and delivery are severely hindered. Leaders of teams in today's world would benefit greatly by investing in communication specialists who can help to bridge the massive gap that is now existing in team communication due to all of the above identified factors.

What can leaders and team members do to help with this goal of stronger communication?

For leaders to better communicate ideas in a video meeting or in chats with their team and for team members to optimally understand and digest information being conveyed, there are several concrete actions that can be taken.

First, everyone needs a lesson in optimizing virtual online communication through lighting. The lighting does not have to be the same as what would be found in a television studio (though it wouldn't hurt), but it most certainly needs to be direct, clear, and without distracting shadows. If a person's face cannot be clearly seen, the effectiveness of the communication is diminished. Take into account that when we believe we are being seen, we pay more attention to the person(s) in front of us and we stay more engaged. Obscured darkness and shadows are not the friend of the corporate leader and her team in the virtual world.

Second, make sure that the sound is top notch. Every room that we are in has its own unique echo. If the computer microphone is being used, don't sit far from the computer as an echo effect can and often does occur. A better option is to use a dedicated microphone that is close to the mouth. Personally, I use a podcasting microphone so that my sound is always optimal. Also, remember not to let the microphone scratch against clothing or other objects. Be sure to speak clearly and enunciate. Too many words and sounds can be lost through virtual communication.

Third, the environment you are in does make a difference. In the virtual world, it is not expected for everyone to be sitting in perfectly appointed corporate settings, but please don't have the laundry hanging behind you or offensive material in sight of the camera. Do not conduct meetings from your messy garage or while moving about the house.

Take into account how you are sitting. Having a dedicated space from which to communicate is ideal. A comfortable, high-backed chair will keep us sitting up straight. Personally, I like a sparsely decorated wall as a background, nothing too ornate or distracting, but also an area where nothing is going to be happening behind me (especially if one has children).

Fourth, eyeline is something that most people don't consider. Our eyeline should be directly aligned with the camera in front of us so that we are not looking up or down, but straight ahead—a perfectly horizontal line from the iris of the eye to the iris of the camera.

Fifth, employ the rule of thirds for the camera. We are conditioned to seeing authority figures on television (newscasters, television hosts, etc.) sitting up straight taking up the middle of the screen with about 25-30 percent of space to camera right, camera left, and above the head. Because of this subconscious conditioning to "pay attention to the speaker," the viewer of online communication has a more "comfortable" feeling when watching or listening to someone who is obeying the rule of thirds. The more comfortable we feel with what we are watching, the easier it is to take in information (i.e., removing distractions from the mind allows more room for new information to be taken in).

Sixth, call out the distractions in the environment. We don't know what we don't know, and if we see someone looking down or looking away, we assume they are not paying attention to us when, in fact, they may simply be referring to notes or checking information or incredibly distracted by something happening in their home. Call it out. If the dog is chewing a cushion, don't ignore it; call it out. Otherwise, the mind remains distracted with the thing in the physical environment, and the person(s) with whom we are speaking question why we are not focused on them and the conversation being had. Call it out.

How can companies' teams be better at collaboration with each other when working remotely?

One of the things that I like to implement when working with teams is a system of repeating information both verbally and in written form after a meeting. It is startling how often a team member will walk out of a meeting and still only have a vague idea of the direction of an initiative or the follow-up action steps to be taken. This is almost always because we all have vastly different communication styles—left brain, right brain, talker, listener, strategist, advocate, etc.

My suggestion is before the meeting ends for each member of the team (provided the team is less than eight people), give a 60-second recap of what they've heard and what they believe the next concrete action steps are. This gives the leader and the other team members the opportunity to hear what was understood (or not), what was reframed in a helpful way (or not), and the opportunity to course correct before actions are taken that lead to confusion and/or workflow redundancies.

Additionally, I like for team members to follow up with leaders and peers with whom they are sharing workflow within 24 hours with a bullet point recap of the takeaways from the meetings, the action items to be implemented, and the timeline for those implementations. This simple, but effective, form of communication can do so very much in helping a team to optimize its communication, build team trust and accountability, and inspire a sense of greater collaboration.

How can these methods be leveraged for video meetings with both team members and clients?

Leveraging these methods can happen by a simple shift in perspective. Through video, it becomes all too easy to be a passive observer of the person speaking, presenting, pitching, or leading. Instead, think of a virtual meeting as a half-hour or hour-long educational children's program on live television. After the program is over, viewers (team members and clients) will remember what they felt from the program (the meeting) more than they will remember the actual words spoken.

Additionally, the viewer (team member and/or client) will only be able to recall what she or he saw and heard on the screen. The screen is limited. The entire environment of the onscreen program can't be seen. The body language can't be read as effectively. Team members will exist in virtual silos. Clients may be left feeling like they've just viewed an infomercial. Leaders must make meetings interactive and participatory. If participants are involved in the process of the meeting by being asked questions or batting about possibilities, they will be more engaged both during and after the videoconference ends.

The television shows that have done this the best are educational shows for children like *Sesame Street* and *The Electric Company*. The children viewing these programs feel that they are included as a part of the show. They learn. They repeat. They expound upon what they've learned. They remember. They think. They feel. They participate and, as a result, feel they are an active part of the meeting, rather than a passive, invisible observer. When we come to think of our virtual world, even in a corporate setting, more like an educational television show where we have a lesson to learn or problem to solve and all viewers (team members and clients) are encouraged to participate and be a part of the show, we create environments of increased learning, value, purpose, appreciation, collaboration, and retention.

It's interesting that the lessons we learned about "learning and growing together" as children we so often overlook as adults. Approaching video communication from this perspective not only makes the meetings more effective but also more fun, leading to greater participant engagement both internally and externally.

What are some ways that an organization can maintain its company culture if it finds itself suddenly all working remotely?

In order to maintain company culture while working remotely, quality communication and inspired collaboration are key. We can become isolated in our home office experience, denied the pleasure and the benefits of small talk, personal interactions, and the quick "check-in" with our teammates. It is helpful to start virtual conversations by checking in with the person to whom we are speaking on a personal level first before diving into work or attempting to obtain the information we need.

Purpose and appreciation, elements of satisfaction in the workplace, are highlighted and reinforced when we feel cared for by our teammates, leaders, and the company itself. During a time when physical interaction is severely limited, more personalized inquiries into a person's health, mental well-being, family, and even upcoming social plans go even further than they do when standing around the water cooler in a shared physical space.

By building communication rapport with the person on the other side of the computer camera, the potential for positive collaboration is increased. We look forward to meeting with a person, interacting, problem-solving, etc.

when we feel engaged with them on a personal level that reflects the core goals and values of company culture. We are people first, having a shared experience of working remotely. When we connect on a level that recognizes this truth, we are much more likely to re-engage with the work at hand and the purpose behind that work.

Additionally, physically separating the workspace from the home space is strongly recommended. Whenever possible, do not take work into the shared living spaces with family members. Work is done in the workspace with work colleagues. Home and family time are for home and family activities. No exceptions! This clear delineation allows our best selves to show up in each space rather than creating a never-ending loop of work, work, work, and ultimately, a feeling of depletion and disengagement.

By observing this boundary, when the time comes to return to work and the culture of the company, a shift in internal gears is experienced, allowing for a subconscious "freshness" with each workday and new activity.

The Watercooler Has Moved—Engagement and Socializing Remotely

It used to be that the watercooler was a place for random, impromptu conversations with employees and managers at the office. If you saw someone heading for the watercooler, you might decide to get a drink and head over to strike up a conversation. The term *watercooler moment* resonates in popular culture. Television shows such as *Game of Thrones*, *Friends*, and *Seinfeld* have been dubbed watercooler shows because people would talk about the previous night's episode around the watercooler with their co-workers.

What this demonstrates is our social nature. We look for a connection with each other, whether it is at work or school or in other settings. Being social and interacting with each other are human nature. It has been observed that when we are not able to be socially interactive, our overall well-being begins to break down in both a mental and physiological manner. People need social interaction in order to remain healthy. A business also needs for its people to have social interaction in order for it to remain healthy as an organization.

I mentioned in the last chapter that a good part of business is built on relationships. That is true in multiple aspects. Building personal relationships with co-workers builds stronger teams. The personal relationships create bonds of trust and a willingness to go the extra mile for a teammate.

193

Good relationships bring good results. Companies should encourage strong relationships between co-workers to build stronger team players. These are virtues and interpersonal skills that will serve them well even after they have left the company. During a forced work-from-home situation, making conscious efforts to enable socialization of the team is even more important. It has positive impacts on the health and mental wellness of your employees. It helps to bring your teams together.

The Impact of Social Isolation

Well before the COVID-19 pandemic, researchers were studying the feelings of isolation that came with working remotely. Even before the additional stress of the pandemic, remote workers had reported feelings of loneliness and isolation. The pandemic compounded that isolation when regional stay-at-home orders were enacted so that going out and meeting with friends after work was not an option. With the added stress of a forced work-from-home situation stemming from a sudden event or crisis, those feelings may also become magnified for your own employees.

It's Far More Than Loneliness

As discussed in Chapter 1, "You Can't Go to the Office: Where Do You Go from Here?", economist Nicholas Bloom studied 1,000 workers from Chinese company Ctrip and found that after nine months of doing their job from home, roughly half of the remote employees would prefer to go back to work at the office. That was true even though it meant a 40-minute commute each way for them. The reason cited by respondents was feeling lonely, but the issues are more complex than a mere lack of company or missing out on spending time with friends. Humans have an inherent need to feel that they belong.

Abraham Maslow was an American Psychologist born in 1908 who was best known for his work on the hierarchy of the five basic human needs that motivate human behavior (Figure 8.1). The base of the hierarchy is a person's physiological needs. Once a person satisfies the needs at that base, motivations for their next higher needs begin to emerge. They build upon each other. For example, the base needs would include items such as food, water, rest, and warmth. Once these needs are met, a person would try to achieve the next higher needs of safety and security. However, safety and security have deeper meanings than merely physical security, and it helps explain the feelings of isolation that emerge from being separated from the team during remote work.

Maslow's Hierarchy of Needs

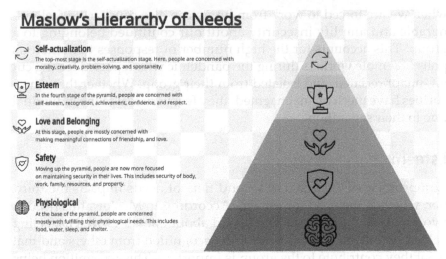

Self-actualization
The top-most stage is the self-actualization stage. Here, people are concerned with morality, creativity, problem-solving, and spontaneity.

Esteem
In the fourth stage of the pyramid, people are concerned with self-esteem, recognition, achievement, confidence, and respect.

Love and Belonging
At this stage, people are mostly concerned with making meaningful connections of friendship, and love.

Safety
Moving up the pyramid, people are now more focused on maintaining security in their lives. This includes security of body, work, family, resources, and property.

Physiological
At the base of the pyramid, people are concerned mostly with fulfilling their physiological needs. This includes food, water, sleep, and shelter.

Figure 8-1: Maslow's hierarchy of needs

Safety and Security

Being in an environment that is familiar and routine gives us a sense of security. It becomes a safe space where we can feel comfortable. We maintain our sense of security and safety when we remain in those predictable situations. Routines such as getting up and dressed, heading out for the commute to work, and walking into the building to go to our workstation Monday through Friday can provide a feeling of security.

We may feel agitated when that predictable pattern is broken, especially if it remains that way for a long period of time. We start to become stressed if we are uncertain when we can return our routine. We begin to feel insecure about our future.

Love and Belonging

The next level of needs involves love and belonging. Again, the motivations behind these needs run deep. We have a fundamental need to feel accepted and be part of social groups. We feel accepted through our positive social interactions with peers. It is the interaction with those groups that help us to experience loving others, being loved, and feeling like we belong.

It's the social interactions that are key to these feelings of love and belonging. Without them, these feelings become less certain and may seem as if they are endangered in some manner. Unknowns that arise when we are forced to work from home can also cause us to feel vulnerable and amplify insecurity about our continued belonging to a group. This accounts for the high number of responses to numerous polls of remote workers during the pandemic who reported feeling cut off, unappreciated, and isolated from their group. Whether these insecurities have merit or are imagined, they feel very real to the individual due to their separation.

Esteem

Compromise of the first and second tiers of needs impacts the third tier, which is the need for esteem. According to Maslow, this need is two pronged. It involves feeling good about oneself as well as feeling valued by others. There is a need for recognition from others and that what they contribute to the group is important. This recognition helps build the individual's self-confidence.

When they do not receive that recognition at regular intervals as they normally would, they feel less confident and their self-esteem drops. At the same time, if the needs below this tier (belonging and security) are not being met, then self-esteem also suffers. Each tier is built on the stability of the level below it.

The Need for Social Interactions

According to the well-known study by social psychologist Roy Baumeister and psychologist Mark Leary, "The Need to Belong: Desire for Interpersonal Attachments as a Fundamental Human Motivation," people need satisfying social interactions with their groups on an ongoing basis with regularity. The need to belong is so powerful and fundamental that a lack of these interactions result in a variety of health issues as well as a desire to form new relationships if a person feels their needs are not being met from their current relationships. In the workplace, this could translate to seeking employment elsewhere.

The lack of social interactions with co-workers leaves many employees with feelings of anxiety, which then magnifies their feelings of being cut off or out of the loop with what's going on with the team. This in turn increases the feeling of being isolated.

The resulting feelings of loneliness and isolation build on each other. Unknowns in any circumstance can make people feel vulnerable. When those unknowns are attached to large aspects of their lives, such as their career paths and their ability to earn an income, it will affect their mental and physical health.

Avoiding Social Isolation

Let's put together all we have discussed about social isolation. A forced work-from-home situation results in an abrupt and sometimes unexpected physical separation from the group. Typical work-from-home employees are separated from the everyday interactions that occur in an office that would ordinarily help reaffirm their feelings of acceptance and belonging. Not being physically around others also means lost opportunity for organic interactions with other members of their group. That loss of daily watercooler moments can negatively affect their feelings of security about belonging to the group. Those negative feelings can become further magnified by uncertainty during a forced work-from-home situation and result in even more stress and distress. As all of the fundamental needs become vulnerable, those in the forced work-from-home situations feel the loneliness and isolation that remote workers have been voicing. They feel disconnected and left out.

Leaving these issues unchecked will result in both physical and mental health issues in remote workers. To alleviate and prevent these negative impacts to employees, managers must alter their own methodologies when it comes to remote team management. Managers must become more engaged and look for signs of stress and possible burnout. Along with deeper engagement, employers must create more opportunities to show recognition and appreciation of team members' achievements. Employers must also create avenues for social interaction between team members. Combined, this all helps team members and their sense of belonging to the group, which is a fundamental human motivation.

In times of high stress or trouble in one's life, belonging becomes a safety net that helps one feel secure again. Belonging allows for opportunities to unload and share with others the causes of anxiety. In business, a strong sense of belonging will help foster environments where people are more productive and overall more confident, dedicated team members.

Be on the Lookout for Isolation

For many companies, the first time they have the entire team working remotely will be when they are hit with a crisis. That means that it may be difficult for those in leadership positions in those companies to know when employees begin to feel isolated and disconnected. It's understandable that managers may not know who on their team may be struggling.

Oftentimes, strong players on the team will choose not to speak up when they are struggling. They tend to try to push through the problem, thinking that the issue is a personal one to overcome on their own. Some of the strongest team members are hesitant to complain, especially during a crisis that affects the whole company.

Those who you see typically as highly connected to the rest of the team will probably be among your most productive under normal circumstances. By highly connected, I mean that they interact with others often and are very involved in projects as well as social conversations. When those individuals feel disconnected, they will likely become less productive as well.

When they feel disconnected, they begin to feel isolated, and that is when they become less engaged with their work and with the rest of the team. Feeling disconnected will lead to feelings of loneliness and isolation, which can lead to employee burnout. At the same time, feelings of loneliness and isolation can lead to feeling disconnected, which can also lead to employee burnout. They are all connected and feed off each other.

What to Look For

A December 2019 article in *Psychology Today*[1] discussed the effects of feeling disconnected in workers:

> *Feeling disconnected from the people you serve often disconnects you from yourself, especially for the many of us who are motivated by a sense of mission and purpose. This disconnection can lead to depression, substance abuse, and even suicide, all well-noted burdens associated with loneliness. So, while loneliness may not in itself be a symptom of burnout syndrome, it is almost universally a consequence. And loneliness can actually cause burnout, as well as one's susceptibility to it. People in the throes of work-related exhaustion, self-doubt, and defeatism are more likely to withdraw, interacting less and less, and effectively isolating themselves from the people around them. And those who already are experiencing loneliness in their lives may lack the emotional and spiritual resources required to feel replenished and resilient under challenging circumstances. A person in this position can be vulnerable to even more burnout, compounding the feelings of isolation and loneliness.*

---
[1] www.psychologytoday.com/us/blog/being-unlonely/201912/
workplace-burnout-and-loneliness-what-you-need-know

In May 2019 the World Health Organization listed Burnout as an "Occupational Phenomenon":

Burn-out is a syndrome conceptualized as resulting from chronic work-place stress that has not been successfully managed. It is characterized by three dimensions:

1. Feelings of energy depletion or exhaustion;
2. Increased mental distance from one's job, or feelings of negativism or cynicism related to one's job;
3. Reduced professional efficacy.

Burnout refers specifically to phenomena in the occupational context and should not be applied to describe experiences in other areas of life.

As you can see, these feelings of loneliness are part of a larger and more serious picture. They are not to be taken lightly or dismissed as something to just power through or ignore. Left unchecked, they can have profound effects on individual team members and on the team as a whole (Figure 8.2).

It's Preventable

The good news is that these pitfalls of remote working are preventable. It is up to managers and team leaders to make sure that they are aware of the signs to look out for and take proactive steps to mitigate potential issues that come with feelings of isolation in remote workers. Employee health is linked to employee productivity, so it benefits companies to foster work environments that enable the well-being of their employees.

Even when a company has a culture in which employee well-being is a top priority, it's still possible for feelings of isolation to creep in during forced work-from-home periods. One must be able to recognize when this happens in order to mitigate the effects. It all begins with being aware of what to look for when you engage with your distributed team.

Changes in Appearance

Keep an eye out for changes in an employee's appearance during video meetings or in-person encounters. Although there may be only a subtle change, if you notice it on a consistent basis, it could be a sign of an employee who is starting to struggle with loneliness, depression, or isolation.

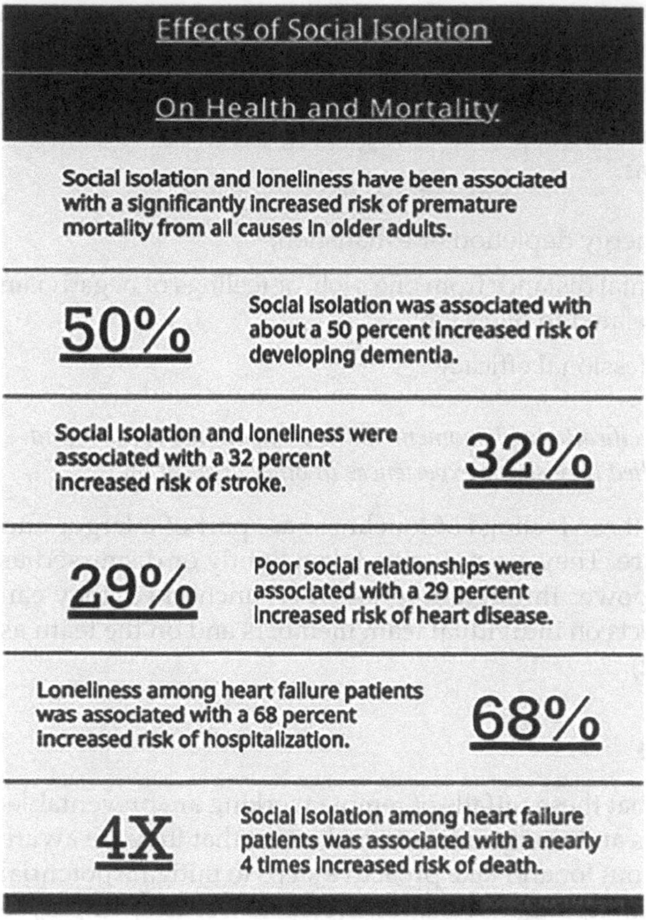

Figure 8-2: Effects of social isolation on health and mortality

A change in an employee's state of mind could manifest as a change in grooming habits or clothing. The employee may wear clothes that are wrinkled or wear the same outfit several days in a row, whereas they normally dress very neatly and put on fresh, clean clothes daily. It could also be a change in the room where they video chat from. A neat and organized workspace may suddenly be strewn with food wrappers, stacks of papers, or unopened mail.

Changes in Communication Habits

You may notice sudden inconsistency in communication habits. Whereas an employee may previously have been very good at responding to questions in emails and chats promptly, they may take hours or even a day or more to get back to teammates. If they stop offering up opinions and input, or stop asking questions to help clarify assignments, they may be feeling less engaged with their work.

Other signs of diminished communication could be that they also stop interacting on social Slack channels and chats with other employees. They may become more unavailable, which results in less collaboration with team members.

It may be a warning sign if they stop turning on their camera during video meetings. This may also be true if you notice a lack of participation in meetings. It may begin gradually at first, but if they stop participating altogether it may be that they are pulling away and are becoming more isolated.

Changes in Quality of Work

Some additional red flags include a decline in the quality of an employee's completed tasks and overall work. Some work may be incomplete or missing details, and they might begin missing deadlines more frequently. You may even see them missing the mark on assignments as if they did not fully comprehend what they were supposed to have been working on. These are all signals of being disengaged.

Cancellations or Lateness

Another sign of being disengaged or isolated is missing out on meetings. An employee may cancel and reschedule a meeting if they are feeling stressed. They may start to skip more meetings that are optional or begin making excuses to skip them.

Their work history may have shown that they are normally very punctual, but then you see them being late for meetings. It may be that they are there on time for the meeting, but because they were running late, they are not really prepared for the meeting. You might even see them take time off without advanced notice or take more personal days then they normally would have when at the office.

Overproductivity

Some employers may think that having overproductive workers is a good thing, but a healthy balance between work and personal lives is necessary for the overall well-being of employees. This balance is key in the prevention of isolation and burnout.

You may see struggling employees start to work much longer hours and be unable to turn themselves off from their work mode. They could be avoiding issues of isolation and disconnection by overcompensating. This could also be true if you find them pulling away from social discussions that they normally would participate in but instead they talk only about work and work-related topics.

High Levels of Distractability

You may see that an employee who is normally engaged during meetings and paying close attention is now more easily distracted. They may be looking at their phones more, sifting through papers, or looking at other browser windows during a meeting. They may also be forgetful or unable to maintain focus during a meeting or while they should be working.

Don't Make Poor Assumptions

Management styles differ, and although some managers are very empathetic to their employees, others are less emotionally aware. A style that lacks empathy is not conducive to remote work, especially during a crisis. It's too easy to confuse the red flags with an employee's lack of care or apathy about their job. Emotionally aware leadership is a crucial component in the remote workplace.

In reality, these performance changes are a warning sign that the employee is struggling and needs help. Reach out to them, but avoid dramatizing the situation or overreacting to these changes when you see an employee becoming disengaged or isolated. You should take action but make it a measured and well-thought-out action instead of a reaction that may add to the stress that you and your employees are already experiencing.

Self-Assessment

Don't forget about yourself. Often, people ignore their own pains and worries when they are in a leadership position. They feel that they need to be a pillar of strength and not show that they too may be having issues. That may be a very noble ideal, but no one can ignore their own health issues and remain effective in their role. Leaders too can experience feelings of being isolated or disconnected while working remotely.

When those at the top ignore their own needs, their employees eventually notice. What's more, when those in leadership begin to struggle with isolation or burnout, they will have a negative impact on the rest of the team and perhaps the rest of the company. It is important not to ignore or dismiss your own struggles. Look out for signs that you are struggling and, if need be, do a self-assessment or an online screening.

Negative Thoughts and Feelings

If you start to feel less engaged and isolated, you may find your thoughts becoming more negative and you may be feeling cynical and more critical of others. You may even find yourself developing a perfectionist out-look and being overly critical when things don't go exactly the way you expected them to. If you don't examine these types of thoughts, they will keep resurfacing. It is best to face them and do some discovery about their origins. Many times, when we are unhappy and we are separated from others, it is easy to invent negative ideas about friends or colleagues or to focus on their traits that we may not view as entirely positive. They may be exactly the same but our own negativity paints them in a bad light. Ask yourself if these things that you are thinking about are actually true and based in fact or are you unintentionally exaggerating them. By examining the truth behind these thoughts, we can assuage these nega-tive ideas and create the opportunity to put them to rest.

If the negative thoughts and feelings are focused on yourself, then ask what you would say to a friend or loved one who was talking about themselves in the same way. Follow the same advice for yourself and confront those thoughts in the same manner. Focus on your strengths and take a break from self-judgment. I find journaling to be very effec-tive. I journal about things that bother me, and as I write, I give myself the opportunity to speak about my feelings and how I am reacting to them. It allows me to get to the source of these thoughts. Once I do, the feelings lose their hold on my thoughts and I can continue my day as I normally would but with a more positive outlook. If the thoughts are pervasive, then consider speaking with someone about them.

Change in Sleep Patterns

Having trouble falling asleep or not being able to stay asleep is a sign that you may be struggling. The occasional bad night of sleep may not indicate a problem. However, insomnia several times a week should make you take notice, especially if you are also experiencing fatigue.

Avoiding stimulants such as caffeine will allow your body to relax better at bedtime. If you must have caffeine, limit the amount and try to avoid it later in the day and in the evenings. Also try to avoid mental stimulants a few hours before bedtime. Working on job-related projects or reading books and articles that will get your mind racing can act as mental stimulants. Instead, read for pleasure and keep to topics that are relaxing.

You also should try to unplug and get your mind and body ready for deeper sleep. Avoid screen time as much as you can prior to bed. Stay off the computer and phone. If you must use them, try an evening setting or a bedtime app on your phone that will help limit what apps are available to you during specific hours that you set and will turn off push notifications during those time periods. With these limitations, you are not tempted to go to social media or other sites that may overstimulate your mind. They will also reduce the screen brightness to minimize the amount of light you see. The blue light emitted by your phone and other screens lowers the production of melatonin in your system. Melatonin is the hormone that regulates your sleep cycle. A reduction in your natural melatonin will make it more difficult for you to fall asleep and wake well rested.

There is a library of sleep apps available for Android phones and iPhones. Many are free or have a membership for full access to all the features. They can help you relax better, track your sleep, or even tell you bedtime stories. Check them out to find one that works best for you.

Fatigue

A feeling of fatigue is a common symptom in extended work-from-home situations. You might find that you can't maintain your energy level as you normally would, or you feel tired and lethargic early on in the day more often and can't seem to shake it.

Even if you have become very good at multitasking, try focusing on one task at a time if you are feeling fatigued. You use more energy to juggle them all. Managing fewer things at once may slow your pace, but it will also reduce your energy output so that you don't feel like you are running on fumes. If you feel like you have less energy, then the goal is to reduce the work that uses that energy.

Remember to take breaks well before you exhaust your personal reserves. Taking breaks earlier in the day will allow you to last longer. The most effective breaks are those during which you do something you enjoy. Try to stick to a daily schedule and log off around the same time each day.

This puts your mind and body into a rhythm and a routine, which in turn will allow your body to better manage your energy levels throughout the day.

Schedule fewer meetings if you've been feeling rundown. The less time that you have to be "on" for a one-on-one or group meeting, the more mental energy you will be able to have in your reserves.

Disinterest

Another symptom you should take into account is losing interest in activities that you usually enjoy. You might push off working on hobbies or avoid some friends and family or social conversations in general. You might find yourself becoming indifferent to activities you used to find fun.

Getting out and enjoying some fresh air and time away from your work can fight the onset of disinterest that can come from isolation. Spending some time in a park or museum or even sitting and listening to music can help you free your mind and relax your thoughts. Go and have some playtime with your pets if you have any, or find another activity that allows you to decompress.

Lowered Motivation

Motivation can also be a struggle when you're feeling isolated. It may become difficult to start or finish work projects because you can't seem to find the strength or interest to get them done. You may even indulge in some productive procrastination and accomplish other tasks to avoid doing the work that needs to get finished.

You can use scheduling, daily routines, and some apps to keep up your productivity. However, being productive and being motivated are not the same. If you start to lose your drive and motivation, the first thing you should know is that you are not alone. Highly successful people have periods when they experience a lack of motivation. It's normal, so do not dwell on it or beat yourself up about it. Instead, recognize it, accept that you are feeling unmotivated, and then do something about it.

Go back and remember why you do what you do. Think about your long-term goals and the people that will come along with you as you succeed in reaching these goals. Take a moment to look at pictures of those people. Think of the higher purpose that you put your effort into and the feelings of accomplishment that you have felt at each step that brings you closer to reaching it. Think of the wins that you have had along the way.

Also, take a look at the work you need to get done and create a plan of action. Break up the work into a series of smaller tasks and put them into an order of operations so that you can knock them off one by one. Your brain will know that you can do those smaller tasks easily. As you look at these more manageable pieces and finish each one, you will gain a feeling of accomplishment and satisfaction that will also help boost your motivation.

Irritability

You might find that you become easily agitated. You may even feel that there is some tension between you and others that you did not have before or sense that others feel tense around you. If left unchecked, you might even let loose some emotional outbursts or feel like you want to.

Try to discover the real source of your irritability and figure out what it is that is really setting you off, and then look to deal with that issue. You may discover that you can easily fix what is upsetting you by addressing it properly. The longer you go without acknowledging the issue and dealing with it, the more it builds up in your system, and it may paint many unrelated things in a negative light.

It is also a good idea to reduce caffeine and other stimulants when you are irritated. Adding stimulants during periods of irritability can cause people to be reactionary and harsh with others. Remember your empathy and compassion and then look to use those traits on yourself. It's okay to be annoyed now and then—it happens to everyone. Just be sure to deal with it. If you have excess energy, a walk with some fresh air and some exercise might be a great way to burn it off.

Trouble Maintaining Focus

You could start to experience difficulty focusing or have a shortened attention span where you cannot maintain your focus. This may also emerge as forgetfulness as well as an inability to pay attention.

Try to get good rest each night and eat well. This will do wonders for your focus as well as your overall health. Too much caffeine and other stimulants can make you feel more scattered, so adjust your intake accordingly. I find that with myself, watching what I put into my body is key to my ability to get a good night's rest and maintain my mental clarity.

You can also add a bit more structure to your day and your tasks. By doing so, you break the day and tasks into smaller, manageable units. Keeping your focus on one task at a time is far easier than multitasking. Consider reducing the number of tasks you take on in a single day and avoid distractions. Turn off push notifications and shut off apps that may pull your attention away from work. Using focus apps that will help you stay off nonproductive apps will be a great add-on in your quest to keep your focus where it needs to be.

Depression

These symptoms may seem similar to those of depression, and truly they are very closely related. Many remote workers who feel disconnected experience depression along with the strains of isolation. Mental health is profoundly impacted by isolation and loneliness.

Feeling disengaged from our group affects our feelings of belonging and being a valued part of the group. Losing that connection can negatively affect our mental health, our emotional health, and our physical health. Should you experience more than one of the symptoms mentioned previously for prolonged periods, follow the advice you would give others you suspect may be suffering from depression. Be sure to speak with your doctor and let them be aware of what you have been experiencing.

Combating Isolation

Remote work does not have to be isolating or result in feeling disconnected from aspects of work or the team. There are steps that can be taken to mitigate the impact of isolation during a forced remote work period. Many of these steps are rooted in company cultures, but a number also involve elements of management and operations.

Structural Changes

Feelings of loneliness are different from feelings of isolation. Many times, feeling as if you are isolated can be due in part to the way remote work is structured. This type of isolation is easier to combat simply by making some changes in operations and workflow.

The goal is to make people, files, data, and general information accessible from anywhere. By doing this, people will not feel cut off from the rest of the organization when working away from it.

Make Work Accessible

Technology has a pivotal role in remote working. Many times, it is limited by the scope in which it is being used. Managers can better enable remote work and reduce feelings of isolation by setting up their teams for success with their tech. Digital assets, documents, and other files should be warehoused in secure, cloud-accessible systems so that people have access to them regardless of whether they are in the office, working from home, or on the road.

Use of a robust customer relationship management (CRM) system will give team members access to client files, their sales history, and support documents to see where there may have been any issues with the clients. This is only the case when CRM systems are being used to their fullest extent. Make sure all departments, from sales to marketing to support, are using the CRM system as intended so that pertinent client data is warehoused centrally within that system. Ensure visibility and transparency by making sure client data and history is available to all teams.

Visibility and Transparency

Teams in different departments should have access to each other across the business. That also means that the info various teams can access should be available across departments as well. Unless there is work that is being done that is classified, there is little reason not to have people fully aware of other projects that different departments are working on. This methodology gives employees a stronger sense of belonging and of being a part of a larger whole.

There is no reason to keep people in a bubble where they interact only with their own teams or have access to knowledge only on their own projects. The more access they have, the more insights they have about their company. This builds a stronger connection to the company and their work. Put files and documents in public folders so anyone from any of the teams can view them. Make project management boards open and not private so that other departments can see them.

This type of visibility helps people see the bigger picture of what the company is doing. This type of transparency in a company culture reduces the feelings of not being in the loop about what's going on at work, especially for remote workers.

I want to point out that I do not mean that you should make sensitive information accessible. For example, HR records and customer payment information should be kept accessible only to those who need such information. However, if sales teams can see the support team's records on customer issues, the sales team can benefit from this knowledge as they groom the clients for future purchases. Likewise, if marketing can see the client notes from the sales team's as well as support records, they can focus future marketing on what features are being used and requested.

Managers and team leaders should also keep themselves visible and available. Open office hours and setting your status to "available" or "online" on collaboration tools and communication apps help people feel that they can come talk to you anytime, even if it's just for a social visit. Simply knowing that you are around and available is good for the remote team member, even if they don't actually need you for anything. There is a comfort in knowing you are there.

Communication

Transparency and the way you communicate are closely related. Remember that it is important to be open and enable people to have access to as much of the company as possible. In many cases, this means that managers have to communicate with their teams more frequently, but that is not a bad thing when done with the right goals in mind.

Be all-inclusive in your communications when possible. Use public channels and chat rooms when reaching out. There are only a few instances where private channels must be used often: when checking in with employees, when providing them with personal feedback, and when giving them one-on-one recognition.

Check-In

Check-ins, catch-ups, and one-on-one communications should happen regularly. Reiterate that you are there and available for your employees. Let them know that they can come and *should* come to you if they need anything or if they are struggling with something.

This repetition is necessary. Those who are more introverted may naturally avoid asking for help and may want to be more private. Those who are more extroverted may also avoid asking for help because they may think that they need to try to push through. All employees need to be reminded that you are there and are willing to let them be heard.

Feedback

Provide regular feedback as part of your communications with remote workers. Receiving feedback on a regular basis gives employees time to take a look at where they've been and where they are going, like intersections on a road map, so they can adjust their trajectrory accordingly to get where they need to be. It lets them appreciate the work they are doing and feel appreciated by those they look up to for guidance. This kind of positive reinforcement can go a long way for someone who feels isolated.

Use the feedback as an opportunity to reemphasize your company values. The values are a big part of your company culture and they are a source of the strength of the company. Point out how the employee's good work and dedication exemplifies company values. Positive feedback is a good way to highlight how these individuals are an essential part of the company and an excellent team member. It enhances their ability to identify as belonging to the group and avoids the feelings of being disconnected or isolated.

Recognition

Showing someone that you recognize their efforts, particularly during trying circumstances, can be one of the best rewards you can give. Feeling appreciated enhances a sense of belonging. It makes them feel like a valued member of the team. Everyone needs a win, especially in a crisis, when small victories can feel big. Take advantage of those small victories. Leverage them to showcase your team and the continued efforts of individual members. Make them feel good about the work they are doing and the sacrifices they may be making as their work and personal lives mesh during their work-from-home period. See them, allow others to see them, and give them recognition.

Make a point to recognize good work and also reasons to celebrate. Celebrate events as if you were all still together in the office. In other words, do the things you would do if someone had a birthday or a work anniversary or closed a big account or finished a huge project. For example, managers can pay for delivery from DoorDash or Postmates to bring everyone cupcakes or other treats to their individual homes so you can all meet up on video for festivities.

Use these celebrations as part of your communication and visibility efforts. Companywide announcements for celebrations are great morale boosters that help people feel connected to each other.

Having a celebration helps with morale, but it is the reasons behind celebrations that are important to keep the group connected. Ignoring those reasons broadens the feeling of being disconnected and distanced. Use the reasons to celebrate and bring everyone together again, even if it has to be virtual.

Mental Health Support

Being open about mental health should be a part of your company culture. That does not mean prying into employees' medical issues, but mental health should be taken as seriously as physical health. If you saw an employee who was looking peaked and feverish, hopefully you would send them home to get rest. If someone seemed unusually stressed or anxious, you should address that in the same manner. Encourage employees to take time off if it seems as if they need it. If you offer medical benefits, make sure they include counseling, if you are able.

Mental Health Toolkits

Create a mental health toolkit. These toolkits are available from many mental health organizations. They offer resources, exercises, and activities to help people who are struggling. You can include local and national resources and place the document in a shared public folder where it is prominently displayed and accessible to all. You could also send out anonymous mental health surveys that help managers get a feel for stress levels and anxiety. This can also help managers take the time to assess their employees and become aware of some who they may not have realized were struggling.

Managers and direct supervisors can positively and negatively influence team members' levels of stress and anxiety. Some employees see their managers as an avatar, or a direct link to the company. If the managers don't seem to care about the workers, the workers will perceive that the company doesn't care either. If managers are empathetic to employees' needs, workers will perceive that the company values them. The style in which they manage people and their workloads can either build stress levels up or de-escalate anxiety.

Encourage Physical Fitness

According to the US National Library of Medicine, exercise has a significant impact on reducing employee stress and the symptoms that lead to burnout. In a study called "Reducing workplace burnout: the relative benefits of cardiovascular and resistance exercise,"[2] researchers found evidence of long-held beliefs of stress reduction from exercise:

After four weeks of exercise participants had greater positive well-being and personal accomplishment, and concomitantly less psychological distress, perceived stress, and emotional exhaustion. Cardiovascular exercise was found to increase well-being and decrease psychological distress, perceived stress, and emotional exhaustion. Resistance training was noticeably effective in increasing well-being and personal accomplishment and to reduce perceived stress.

Many companies provide gym memberships or reimburse employees for exercise classes. Doing so is good for the company. It keeps the team in better health, and it reinforces the company values of caring about the team members. It promotes physical health, which in turn promotes better mental and emotional health for all.

Engagement

Many companies have a company culture that values engagement and intentional connections. This means that managers and others make it a point to reach out to others within the company to create authentic connections beyond the traditional employee/employer transactional relationships. As part of everyday work, this type of culture can avoid the strains of isolation for remote work. It is part of a larger business strategy that keeps the company and the employees healthy as well as happy. Let's take a look at the importance of employee engagement.

According to Gallup, in their report "What Is Employee Engagement and How Do You Improve It?" there are big differences between highly engaged employees and workers who are not engaged with the company. It showed that engagement also drives retention and productivity.

> **BEST PRACTICES**
>
> Engagement should be aligned with other company values. It should be part of the everyday workplace where employees feel valued because they are valued.

[2] https://pubmed.ncbi.nlm.nih.gov/25870778/

Engaged employees are highly involved in and enthusiastic about their work and workplace. They have psychological ownership in the company. They drive performance and innovation, and move the organization forward.

Employees who are not engaged are psychologically unattached to their work and company. Because their engagement needs are not being fully met, they're putting time—but not energy or passion—into their work.

Actively disengaged employees aren't just unhappy at work—they are resentful that their needs aren't being met and are acting out their unhappiness. Every day, these workers potentially undermine what their engaged co-workers accomplish.

Showing Up and Staying

Engaged employees make it a point to show up to work and do more work. Data supports this fact, as is shown in the Gallup Workplace study "The Right Culture: Not Just About Employee Satisfaction": Highly engaged business units realize a 41 percent reduction in absenteeism and a 17 percent increase in productivity. The Gallup study goes on to say the following:

Engaged workers also are more likely to stay with their employers. In high-turnover organizations, highly engaged business units achieve 24% lower turnover. In low-turnover organizations, the gains are even more dramatic:

Highly engaged business units achieve 59% lower turnover. High-turnover organizations are those with more than 40% annualized turnover, and low-turnover organizations are those with 40% or lower annualized turnover.

This shows that engagement drives the connection between the employee and the company as a whole. As the connection strengthens, so does the sense of belonging. Team members have a deeper commitment to the entity when they feel a part of it, which in turn drives their performance for that community.

When workers are forced to be apart from their group or community, the engagement must continue. In fact, to stave off the feelings of isolation and loneliness, engagement must not only continue, but it must be ramped up to make up for the lack of in-person connections. Managers must find ways to engage remotely so that the employees still feel as connected as they did when their group was all together.

Not Out of Mind

Just because your remote workers are out of sight, they should never feel out of your mind. This is where engagement can help them feel connected with the company and the work they do.

Soliciting opinions from your employees can help them feel valued and like an integral part of the group. In general, people only ask opinions of people whose thoughts they value and respect. An especially useful application of this concept is to ask for opinions after projects are completed. Managers can use the insights from team members who work on those projects to set up future projects in an optimal way. It creates an opportunity for engagement that benefits the whole company.

Engagement Aligned with Other Values

Engagement should be a part of the company culture and a strong part of the company values. For engagement to be effective, it has to be authentic and consistent—meaning not just on a regular basis but consistent in how the company does day-to-day business and how it treats employees. It should not be an add-on or something that is done as a reaction to teams feeling isolated. If you don't currently have employee engagement as a top priority, then it is important to add it immediately.

Foster Openness

Part of employee engagement is reaching out, but managers should create an environment where employees feel comfortable talking to them as well. Employees should feel they can come to you should their workload become unmanageable or if they are struggling with stress.

Being available to employees is a strong part of creating an environment of openness. Again, regular open office hours and showing yourself as available on chat channels or other collaboration apps is like having an open door. Just being there to listen can be a huge help in combating feelings of isolation.

Listening Actively and Empathetically

You may be familiar with the concept of active listening. It is a communication technique in which you listen closely to the other person and take note of nonverbal cues as well as respond in a way that shows that you both heard and understood what the other person was conveying.

It is a great tool for effective workplace engagement. Empathetic listening is also a great skill to have when in a leadership position. Empathetic listening involves listening and responding in a way that shows that you understand the message and recognize the *emotion* behind the message. It helps validate what the other person is saying so that their words are being heard and what they are feeling is also being recognized. Both of these skills help build connections with employees and further engagement.

Social Needs

One of the big realizations from the COVID-19 pandemic is that humans are social beings who need to regularly spend time with their groups. Without that interaction, we become stressed and anxious. Spending time interacting in-person with one's group has a profound positive impact on emotional and mental health and an individual's sense of belonging.

Since working from home means that people are cut off from the unplanned interactions that happen in the workplace, it is important to plan these social interactions and create opportunities for interactions among your team members. The organic in-person interactions must be replaced with more intentional interactions, and manufacturing spaces where more natural forms of interactions can happen must also be a part of socializing strategy.

Creating Non-Work Spaces

As a company modifies the way it works during a time of remote work, the company must also modify the way it spends time together. In-person norms won't apply. The company still must work and socialize together if the company is to thrive during the crisis that forces you to go fully remote. As you create rooms, chat channels, and other virtual spaces for work, you should also create similar spaces for casual discussions about non-work-related topics.

Many companies have created Slack channels or other chat rooms that are for socialization. You might even dedicate a channel to a specific topic, such as "what's for dinner," where people exchange meal ideas and recipes. Spaces for sharing song playlists from Spotify or SoundCloud are also very popular. Even spaces dedicated to discussing television shows just like many would do around the watercooler has become a remote workplace norm.

Get the team involved in creating these spaces. Solicit their opinions, get feedback, or allow them to create the spaces on their own and have some fun with it. What's important is that these virtual spaces are created to empower the team to have real interactions with each other.

Virtual Lunches, Coffee Breaks, and Happy Hours

Another way to maintain the social aspects of a workplace is to plan virtual moments of togetherness that mimic real-life events, such as going out to lunch or coffee, or getting together for a drink after work. Working remotely should not mean that these events no longer happen. If you are not able to do them in person, then do them virtually.

During the shutdowns of the pandemic, many managers and team leaders hosted social meetings with their teams where they all shared a drink or meal. To add to the social interaction, team leaders might send boxed lunches or breakfasts to each employee from a delivery service so it would feel more like they had all gone out to eat together.

These small acts of thoughtfulness and detail go a long way. One friend of mine received a small bottle of premium vodka, a mixer, and a fresh lime, along directions to mix specific cocktails for an online happy hour from a vendor that would have normally taken his team out for drinks when the salesperson was in town.

There are many creative ways to make these virtual social events engaging, meaningful, and authentic so that people can maintain their sense of belonging.

Make Pre-Meeting Time

If you plan for there to be some time prior to the start of the meeting for co-workers to socialize, they will more fully enjoy the rest of the meeting, and they are more likely to stay engaged. This socialization would not be appropriate for every meeting, but if you take the first five minutes of a Monday morning check-in for free interactions, you will see a boost in morale. Those times, although structured, will still create a regular space where people look forward to sharing news from the weekend and hearing their co-workers' stories.

It's also a natural opportunity for managers and team leaders to interact as team members, not as someone higher up in the organizational hierarchy. These pre-meetings can be additional opportunities to foster an environment and culture of openness.

Celebrations

As we discussed in the context of recognition, celebrations are important to us as social beings. It's not the party or the cake and party themes that are important, it's the reason behind the celebration that is critical to us: the recognition of milestones and accomplishments.

Those are important in any group, but they are especially important when we are forced apart from the group. Not celebrating milestones and accomplishments would only add to a sense of grieving for losing the normalcy that would come with those occasions. Such celebrations drive the sense of inclusion and improve employee morale when they are recognized or see their co-workers recognized.

Encouraging Friendly Competition

Many companies have created some friendly and sometimes silly competitions with their groups to keep them connected. One example is posting photos of meals the employees created over the weekend. Other companies might have co-workers decorate their video work spaces at home with a theme. Some might have fitness challenges in which they would count steps on a given day.

Winners could be rewarded with gift cards or food baskets. The contests and prizes themselves are not as important as the effect they have on the distributed team. These small competitive games create some fun and give everyone a feeling of togetherness.

Plan In-Person Events

If at all possible, plan in-person events for the team. This is an important way to foster engagement and to combat isolation. In-person events build authentic bonds and improve existing bonds between people. These events could be lunch, happy hour, or even meeting up at a park. An outside group meeting where you share some snacks can go a long way for team members who miss working side by side. Be conscious of people who may have long commutes. Pick a spot that is easy for everyone to get to so everyone can be there.

Company Values and Mission

Companies that are successful with engagement are the ones that have the well-being and importance of the employees as part of their values and company mission. In other words, a foundation of the company is the recognition that the team members are vital components to the success of the company and they are treated as such.

The intent for fostering engagement should not be to improve productivity. Instead, a company should strive to ensure that their employees are healthy and fulfilled and feel that they belong and are a part of something great. Reaching that goal improves productivity.

Set the right goal, and keep pushing to meet it. That will create the right work environment for authentic and honest engagement with your teams, regardless of whether they are in person or miles apart.

People First

As humans, we all have distinct needs, and as team members we depend on each other for many of those needs. Being social is indeed a need for human beings. It's not just an afterthought or optional. Company leaders need to come to the realization that these social interactions for remote teams are as important to the job as having the right equipment to work. You cannot maintain a healthy and cohesive team without them.

The following interview involves a business leader who understands the importance of that sentiment. Taking care of your people will ensure that they are willing and able to take care of their group (the company) and the clients.

BRIAN HANDRIGAN, CO-FOUNDER AND CEO ADVOCADO

Company Profile

- Location: St. Louis, MO
- Employees: 17
- Primary Line of Business: We bridge the gap between offline audiences and digital behavior. We've built a data platform that tracks and coordinates multiscreen campaigns in real time.
- Primary Audience: We arm brands, media companies, and agencies with data that helps them target their multiscreen campaigns more effectively and maximize ROI.

About Us

Advocado is a cross-media managed data platform that instantly generates, integrates, analyzes, and activates data. The platform fills visibility gaps with data that brands, media companies, and agencies can't get anywhere else, helping them target their multiscreen campaigns more effectively, maximize ROI, and bridge the space between offline audiences and digital behavior.

What tech services/software did you use to go remote?

Like many companies, we've leaned heavily on applications such as Zoom, Slack, and Salesforce.com. We're using LinkedIn much more to stay connected to what's going on in the industry and be part of conversations there.

What was the most difficult part of going fully remote for you and your team?

We place a ton of emphasis on our company culture here at Advocado, so when we transitioned to working remotely, the only shift was our geography. We start each day with a 15-minute morning meeting that we like to call "Huddle." It's our time to check in with a daily gratitude, discuss what's on our schedules for the next 24 hours, and call out any items we're stuck on that need another team member's attention. We've found that meeting each morning sets the tone for the day, as well as keeps everyone connected. It also cuts down on excessive meetings, emails, etc. It's a practice that has been part of Advocado's company culture even before going remote.

What was the easiest part of going remote for you and your team?

We never lost the core of who we are as a company. Our mission is to inspire others to make their dent in the universe, and our company culture is the biggest part of who we are. It's truly the heart of the company. We're constantly in contact with one another, with each team coming together weekly to discuss professional goals as well as personal goals too. Some of the employees we've brought on in the past year haven't met their colleagues in person yet, but we've still been able to function as a family and get to know one another.

How did everyone working remotely affect your team working together?

It's only strengthened our company as a whole. While we aren't physically together, team members are always checking in, collaborating across departments on projects, and sending messages via Slack channels.

Did you notice much difference in how your team worked together when remote?

This was definitely a fear of ours, but in reality our efficiency may have even improved. As a company that practices Traction EOS, our weekly 90-min Level10 meetings for all departments and the executive team keep everyone communicating and problem-solving in a collaborative manner.

Did you create any avenues or methods for your people to stay social with each other?

We use Slack extensively here. While we have channels for our business operations and specific collaborative efforts, we also have chats like #random, where we've gotten to know each other's pets, family life, and what's going on in each other's lives outside of work. We've had fun playing around with Zoom video filters during holidays, or just using filters to make team members laugh during our daily Huddles. Outside of technology, it's just who we are to congratulate and recognize team members for the work they've done, or the milestones they've reached in their personal lives. We've hosted happy hours and played trivia as well.

How did you work remotely with clients?

Virtual work has enabled us to do more, because there hasn't been any travel involved. Before, we'd schedule business trips across the country that were months away, but now we're able to take calls with clients much quicker. It's helped us further relationships with contacts that we had before working virtually, and get to know new prospects as well.

What would you do differently?

Process-wise, I'm not sure there is much we would change since it all went so smoothly. However, looking ahead, if we had to do it again, I would want to negotiate terms in our lease for creative rent relief and would have a better equipment management system and means for setting up affordable yet functional remote work spaces for our team.

What advice would you give?

The biggest advice I would give is that successful remote work is less about working remotely and more about creating a team-first, high-trust company culture that can adapt to physical location changes when needed. WFH was so successful for us because we had a high-functioning and high-trust team in the first place. When everyone on the team is genuinely committed to helping each other be their best selves on a daily basis, everything else is just kind of easy to deal with.

Generational Struggles

We live in a time in which several generations interact within the workforce. In this chapter we discuss some of the traits of this generational workforce and we will discuss the way these generations work as part of a newly distributed team. In a time of crisis or abrupt change, people from diverse generations may react differently to the circumstances based on their world view and what they value in both their personal and professional lives. I want to be clear that it is important that we do not make assumptions that people within any specific generation will all act in a particular way when faced with the same situations. That thought process leads to stereotyping and use of blanket terms, which we should avoid.

Humans are far more complex than simply being a product of their generation. However, members of each generation share commonalities of world events that impacted their youth, which have a level of influence on their outlooks. Additionally, the current age and stage of life will shape the view of what is important to an individual in terms of their personal lives and careers. A 20-year-old person will most likely have dissimilar criteria for job searches and be at a different point in their career path than a 50-year-old person in the same industry. Although we will be speaking in general terms about the generations, it's important to keep in mind that as unique individuals, your employees won't all react in the same ways as others from their own generation. The content of this chapter will provide you with an overall framework for understanding generational differences so that you can be sensitive to any generational struggles you may see in your employees.

As we have discussed, many companies during the COVID-19 pandemic did not have a plan for an emergency in which their business had to suddenly work remotely. This means that, practically overnight, an intergenerational workforce had to leave their offices and move into an uncertain environment where they all needed to adapt at the same rate. It became clear that, in general, the different generations reacted differently to their new situation, and they did not necessarily react in the ways many would have predicted.

These generational differences make sense when one considers that they each have different needs and priorities. They are not only at different stages in life, but the workplace has changed dramatically since each generation entered it. The work culture for baby boomers entering the workforce was very different from that of Gen Z as they enter the workforce today. Additionally, baby boomers are looking toward retirement while Gen Z is looking toward climbing their career ladder. These various points of view dictate what each group sees as important in their job and in how they believe work should be done.

If managers and team leaders want to have the company continue to thrive when the team is distributed, they must understand the varying needs of this intergenerational workforce. They must take into account the needs and concerns that each generation prioritizes. This deeper understanding will enable leaders to adapt their management styles to effectively engage, communicate, and collaborate with them all.

The Generations

Generational perspectives play a role in how people view and respond to a crisis or change in general. Having multiple generations in any single workplace means that employees will have different outlooks and adapt differently to new work situations.

Managers must avoid using this information to stereotype their team with different generational labels. Instead, they should view it as adjusting their managerial methods to be more inclusive by taking all priorities and needs into consideration. This kind of empathy and deeper understanding leads to more authentic engagement. With that in mind, let's take a look at these demographics. This generational information was originally published by Perdue University as part of its Education Partnership program.[1]

Baby Boomer

Born: 1946–1964; 10,000 baby boomers reach retirement age every day.

Traits: Optimistic, competitive, workaholic, team-oriented.

Shaped by: Vietnam War, Civil Rights Movement, Watergate.

Motivated by: Company loyalty, teamwork, duty.

Communication style: Whatever is most efficient, including phone calls and face-to-face conversations.

Worldview: Achievement comes after paying one's dues; sacrifice for success.

Gen X

Born: 1965–1980; by 2028, Gen Xers will outnumber baby boomers.

Traits: Flexible, informal, skeptical, independent.

Shaped by: The AIDS epidemic, the fall of the Berlin Wall, the dot-com boom.

Motivated by: Diversity, work-life balance, personal-professional interests rather than the company's interest.

Communication style: Whatever is most efficient, including phone calls, face-to-face conversations, and video calls.

Worldview: Favoring diversity; quick to move on if their employer fails to meet their needs, resistant to change at work if it affects their personal lives.

Millennial

Born: 1981–2000; by 2025, millennials will make up 75 percent of the global workforce.

Traits: Competitive, civic- and open-minded, achievement-oriented.

Shaped by: Columbine, 9/11, the Internet

Motivated by: Responsibility, the quality of their manager, unique work experiences.

Communication style: Instant messages, texts, email.

Worldview: Seeking challenge, growth, and development; a fun work life and work-life balance; likely to leave an organization if they don't like change.

Gen Z

Born: 2001–2020; Gen Z has never known life without Internet-connected smartphones.

Trait: Global, entrepreneurial, progressive, less focused.

Shaped by: Life after 9/11, the Great Recession, access to technology from a young age.

Motivated by: Diversity, personalization, individuality, creativity.

Communication style: Instant messages, texts, social media.

Worldview: Self-identify as digital device addicts; value independence and individuality; prefer to work with Millennial managers, innovative co-workers, and new technologies.

Figure 9.1: Birth years of four primary generations in the workforce

Figure 9.1 offer a quick snapshot of the birth years for each generation. As you can see from the preceding information provided by Purdue University, the influences on each of these four groups are very different world-shaping events that occurred during their childhood and teen years. Those influences help shape their world view and also impact what is important to them when it comes to their career paths and their motivations when it comes to the companies that they work for.

We should also take into consideration the changes in the management styles and workplace attitudes in how employees have been treated over the decades. Someone who entered the workforce in the 1950s would have been immersed in a very different set of corporate values and work cultures than a person who enters the workforce in 2021. A good example would be the differences in the workplaces shown in television shows such as *Mad Men* and *The Office*. *Mad Men* depicted the workplace place of the 1960s and *The Office* took place during the early 2000s. Using these two examples and the corporate cultures of those eras, it's easy to see that what was considered acceptable in the workplace in the 1960s would be a nightmare for an HR professional of today. Today, companies are shifting to a more employee-centric company culture, where there is commitment to the employee, inclusiveness, and open communication. This is a big shift from decades past where the annual review was the main guide for feedback and judgment of an employee's performance.

The generational influences and work cultures combined with their stages in life shape what they value and what they perceive their needs to be. This impacts everything from how they communicate to what they see as real benefits in a job.

Each group has different strengths and different needs when it comes to work, yet these differences are not as predictable as one may think. There is a wide range of ways these groups adapt to sudden change and react to times of crisis. Many had predicted that baby boomers in the workforce would have had the most difficulty adapting to the sudden and prolonged work-from-home situations during the pandemic. However, a 2020 workplace study by Oracle showed that it was Gen Z who had the highest rate of burnout. This study echoes multiple other surveys and studies that show that Gen Z and millennials were more negatively impacted by stress and anxiety at work during the pandemic than other generations in the workforce.

Managers must have a deeper understanding of how these different groups respond to abrupt change when there is a sudden switch to remote work. Managers must adapt their style of management as well as engagement strategies to meet the needs of the team. It cannot be a one-size-fits-all methodology if it is to be authentic as well as effective.

There Is No Generational Gap at Work

Most employees share the same pain points and have more in common than not. People in each generation want to feel appreciated by their co-workers and their employers. They want to work in a healthy environment where their needs are met and their work is valued. They want to feel secure in their job position and in their future. However, each group has its own unique challenges and priorities. Consider the general differences in home life among generations. Some have recently become empty nesters, others are having their first child, and still others are thinking about getting their first solo apartments. These different living situations will also impact the merging of the work and home lives during forced work-from-home time. These generational workers may share similar values, but they have different priorities due to these life stage challenges in their home and how far along they are on their career paths.

Negative Stereotypes

Sometimes other employees and even managers will pigeonhole some of their team members by falling for negative stereotypes about the various generations. Some may think that baby boomers are technophobic or resistant to new methodologies or new ways to do their jobs. Some may believe that millennials may act entitled or that Gen Z is too dependent on technology. Such labeling and stereotyping damages the connectedness to the group and can make people feel further isolated. Even if such thoughts are not stated out loud, they can manifest themselves in assumptions a manager may make about the work, behaviors, or attitudes of those they supervise.

Managers and company leaders should discourage anyone from buying into these stereotypes. Yes, there are differences in each generation. However, the individual team members are far more complex than simply being a product of their generation.

Further, the differences the generations have are not gaps of work ethics or gaps in the ability to do the job. When leveraged correctly, the differing viewpoints and priorities are a bridge to a stronger team. Each generation of worker contributes a wide range of strengths and methodologies to get the job done.

Work Styles

By taking a strengths-based approach to multigenerational teams, you can maintain productivity and keep the teams engaged during your full remote periods and well afterward. The better you understand what motivates your team, the better you can guide them to reaching mutual goals for themselves and the company. Understanding your workforce can also help you with goals of more effective communication and engagement.

Priorities

The key to effectively engaging with different generations is to understand their priorities both in their home life and their professional lives. Nintex, a process management and automation company, published its 2021 Workplace Study, which surveyed 1,000 US-based full-time workers at companies with 501 to 50,000 employees. This study showed notable differences in workers from the various generations during the sudden switch to working from home in 2020. The study made it clear that there was a definite distinction in what each generation wanted in order to make their WFH experience better. Some of these differences are shown in Figure 9.2.

"What Would Make Remote Work Better?"

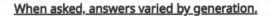

When asked, answers varied by generation.

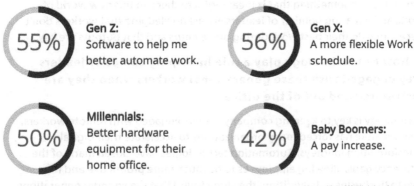

Gen Z:
55% Software to help me better automate work.

Gen X:
56% A more flexible Work schedule.

Millennials:
50% Better hardware equipment for their home office.

Baby Boomers:
42% A pay increase.

Source: Nintex 2021 Workplace Survey

Figure 9.2: Survey: What would make remote work better?

BEST PRACTICES

Discourage negative stereotyping of generations on the team. Managers and company leaders should focus on commonalties within the group while also understanding individual challenges and priorities.

I spoke with Terry Simpson, senior solutions engineer at Nintex, and asked him some questions to dive deeper into the study. The interview appears in the sidebar "Generational Differences in Remote Work."

GENERATIONAL DIFFERENCES IN REMOTE WORK

1. There seems to be a difference in how Gen Z, millennials, Gen X, and baby boomers are affected by a sudden shift to remote work. What are some action items that company leaders can take to make sure that everyone on their teams can stay productive and yet avoid added stresses that can lead to burnout?

In the study we noticed that very different needs for each generation became apparent, so company leaders will need several different options or considerations when responding to employee well-being. Respondents identified four primary needs that were all key to managing a healthy remote working experience: better hardware and software, more flexibility, mental health resources, and more time. Each generation responded differently to each of these needs. An example of this was with baby boomers and Gen Z: Baby boomers were more financially motivated compared to Gen Z, who were more process driven and expressed hardware and software needs. Leaders can respond by ensuring their teams have access to the technology they need to work effectively, freeing them up to focus on less mundane tasks and to take time out to do something that is meaningful to them. In this new world of work, an increasing number of leaders are acknowledging that workers don't necessarily do their best work during office hours and that outputs are key.

2. How can technology play a role in helping company leaders stay engaged with these generational workers when they are distributed and out of the office?

Technology is key to keeping company leaders engaged with remote workers. Being remote has forced many organizations to accelerate their digital transformation journeys. Automation technologies have closed many of the efficiency gaps, allowing employees to be much more productive and responsive to the business. In addition, the study found that the younger generations were more demanding of having great technology incorporated into their job functions. In many cases, younger workers have always had the Internet and software platforms integrated into their daily lives. Working for an employer who has outdated processes makes for a less efficient and less engaging experience for those workers who are accustomed to the speed and ease-of-use on offer from technology solutions.

3. Do you have any additional advice or words of wisdom that you can share about the topic?

Employees haven't just adapted to fully remote work—the majority are doing well in this new environment. Seventy percent of respondents said their experiences working remotely during COVID-19 have been better and more productive than they expected. When asked to describe their better-than-expected experiences, respondents pointed to family time, no commute, fewer interruptions, and work-life balance.

Not all WFH experiences are created equal. The experience of working remotely is correlated with job level: The more senior you are, the more likely you are to adapt to remote work and report higher productivity. Entry-level roles are experiencing overwork, confusion, and living situation challenges more so than their more experienced colleagues. Overall, employees reported positive remote work experiences. But employers should acknowledge that individual experiences vary, and some employees may need more support or flexibility.

Keep in mind that this sudden shift to remote work fundamentally changes how each group is doing their work and how it impacts their daily lives at home. By focusing on the various needs and priorities of each group, managers can effectively mitigate the challenges that each group will face. This in turn increases productivity and maintains a work environment that keeps everyone feeling fully engaged with the company even during times of crisis or abrupt change. Job satisfaction and a healthy work-life balance are part of the equation.

Before we look deeper at these generations, take a moment to look at the population of each group (Figure 9.3). Take into account not only their proportion in the workforce but as consumers and as part of the community. Each generation is uniquely positioned to influence how work will be done and how the business world will adapt to meet their needs both as workers and consumers.

BEST PRACTICES

A higher level of automation, up-to-date processes, and the right technology will create a better work-from-home experience for all remote workers. Review your process and your tech to ensure that they are updated for the virtual office and a distributed team.

Baby Boomers

Among these four groups, baby boomers are the generation that have spent the longest amount of time in the workforce. Therefore, their career goals are very different from those of Gen Z, who are just entering the workforce. For boomers, their goals are geared toward saving for retirement, health benefits, and salary levels. This group has a tendency to put their work life first due to a strong sense of duty and their appreciation of a strong work ethic.

The Population of The United States by Generation in 2019

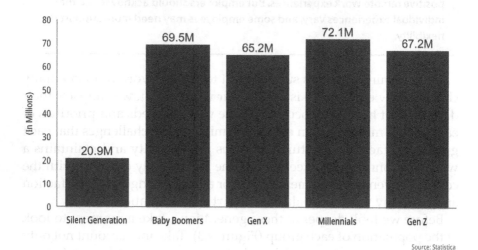

Source: Statistica

Figure 9.3: The population of the United States by generation in 2019

This group also reported the least amount of trouble adapting to the sudden WFH shift during the pandemic. They had the lowest instances of burnout of any generation in the workforce. This speaks to their ability to adapt. That means that abrupt changes had less of a negative impact on their ability to switch off for the day and maintain a healthy balance between their work life and personal life even when working from home.

Boomers tend to be higher up in companies when it comes to position and salary. This may be why some of the younger generations believe that the boomers have all the good jobs. At the same time, baby boomers are well entrenched in the gig economy. Many are consultants or independent contractors. They want to keep earning even after retirement. To do this, they must remain competitive, so they are often well versed in remote technologies and will adapt to new tech to get the job done. They have a desire to continue to be a strong and capable part of the workforce. Managers should provide on-the-job training for learning new roles in the company and opportunities to adopt new technologies.

Because of the length of time they have spent in the workforce and their experience, boomers tend to be confident decision makers and are well focused. Their work tends to be process-oriented, team-oriented, and goal-oriented. Boomers collaborate well with peers regardless of generational differences. Boomers are also community minded and philanthropic individuals who desire stability and are loyal to employers who treat them well.

As noted by the fact that many boomers are strong players in the gig economy, they are seasoned upskillers. Upskilling is the practice of improving a worker's ability and skills by providing added training and opportunities for professional development. This eliminates skill gaps and makes the worker a more valuable employee. They are happy to learn new skill sets and new methodologies. Boomers as a whole also prefer flexible work schedules, which give them a better ability to manage their time. This may also be why this group has a stronger work-life balance, and that balance is likely the reason for their low rate of burnout.

Generation X

According to a Pew Research report titled "Generation X: America's neglected 'middle child,'" Gen X seems in many aspects to be the overlooked generation. Certainly, this generation does not get the same type of media attention millennials and boomers get. There are countless articles dedicated to millennials and boomers, but reporters and bloggers often neglect the generation sandwiched between the two. Further, baby boomers and millennials are both larger generations, which makes the Gen Xers the minority in the workforce.

Gen Xers are the children of the hardworking boomer generation. Many people in Generation X saw their parents as workaholics who prioritized work over personal lives. Many of their parents suffered from varying degrees of burnout due to their hard-work ethic. This could be why Generation X is credited with pushing the idea of a better work-life balance. Many Gen Xers have focused on their children and enjoy flexibility to allow them to spend time with their families.

This generation is eager to advance in their companies. They tend to have well-defined goals and prefer to have clear expectations about their job and tasks. They prefer to have open conversations and tend to be direct in their communications. Many are nearing or hitting their peak in their career path.

Public access to the Internet was new when many in Gen X were still teens. Having a computer in the home while they were growing up was far less common than it is today. Internet access came in the form of cludgy, slow dial-up connections that were tethered via a phone line to a wall jack. In truth, the computer revolution was still in its early stages when Gen Xers were children. They were young adults when mobile phones became popular and then developed into Internet-connected smartphones a short time later. Because smartphones became commonly used when Gen Xers were adults, this group may not be considered by some to be as tech savvy as their younger counterparts. However, as Gen X did not grow up texting or using social media, they are considered to be strong communicators who are able to more easily engage in conversations and easily make genuine connections with clients and co-workers. This also makes them great independent contractors who excel at being team players.

A *Harvard Business Review* report called "A Survey of 19 Countries Shows How Generations X, Y, and Z Are—and Aren't—Different" highlighted some of the attributes of Gen Xers. It showed that Gen Xers make great leaders, with goals of not just looking forward to added responsibility and the opportunities to lead, but also looking forward to coaching and mentorship. Many people in this generation have 20 years of experience in their current fields of expertise. Generation X has the most work experience after baby boomers, among members of other generations.

Gen Xers have a high rate of entrepreneurship, likely because they grew up during the early 1980s recessions and high inflation. Many of their parents, who had come from a work culture in which many people stayed with a company until retirement, got laid off during that recession. This caused Gen Xers to develop a culture of distrust with large corporations when it comes to their financial future. Gen Xers do not want to go through the same financial hardships as their parents.

This group is sometimes referred to as the "latch-key generation" because they grew up at a time when there was a high rate of divorce and when both parents worked outside of the home. Many times, these kids were left alone at home and would be told to lock the doors when they got home from school and not to open them up for strangers. Because of this, members of Generation X are very independent and prefer not to have a great deal of supervision. They will get the job done if you communicate clear expectations and are honest in your discussions.

This independence could be a big reason why Gen Xers tend to be very innovative and forward thinking. Some of the more disruptive brands that were founded by Gen Xers showcase these traits: Twitter, PayPal, and Google, to name a few.

Millennials

In 1987, William Strauss and Neil Howe, fathers of the Strauss–Howe generational theory, came up with the name *millennials* to describe the generation of individuals who would start to graduate from high school in the year 2000. Many of this generation were new to the workforce around the time of the Great Recession that took place in 2007–2009. Millennials have been dubbed the first generation to be worse off than their parents. The recovery rate for employment of millennials after the Great Recession was slower than it was for older generations. Fewer millennials owned homes at the age of 25 to 30 than previous generations. They are the first generation to not out-perform their parents' salaries by the same age and also tend to have a higher student loan debt than Gen X or baby boomers.

Millennials are very community minded. They want to drive social change. We can even see their desire for social change in their purchasing habits. Many millennials will switch brands based on the causes a company supports. They may boycott a company because the company donates money to a cause that does not align with their beliefs. This shows that this generation wants to feel that their actions in daily life are making a difference. It also shows that they have a desire to be part of something bigger that has a positive impact. They are focused on the higher good. This means if they are well engaged with a company, they will work hard to take care of their group.

This is where company values and the brand's mission should be emphasized as strong parts of the company culture when engaging with millennials. Company ethics and social responsibility are important values to them. They want to achieve great things, and they would love to work with a company in which they can achieve that goal and help make a difference.

Regular feedback from company leaders is paramount for millennials. It helps them know how they are doing, gives them clear direction, and keeps them engaged. Unlike the older generations, millennials grew up using emerging forms of digital communication, such as chat and text messaging. They have sometimes been referred to as the oversharing generation that posts their lives online, unlike older generations. Because they tend to share so much, they are very open in their communications. This style of sharing makes them accustomed to feedback on a consistent basis. If they do not receive it, many tend to lose interest in what they are doing.

Millennials are unafraid to switch jobs and to enter new territories. A 2021 Gallup poll called "Millennials: The Job-Hopping Generation" showed that 6 in 10 Millennials are open to new job opportunities. One reason this generation has a tendency to switch jobs frequently during their career is a lack of engagement from the company. Employers should look to give millennials opportunities for longer-term career growth where they can continue to be vital parts of the organization in new ways. If management does not keep in mind what drives and motivates millennials, millennials will feel a lack of authentic engagement, which leads to a lack of attachment to the company or their job. Here are some additional statistics from the same Gallup Poll.

Gallup has found that only 29% of millennials are engaged at work, meaning only about three in 10 are emotionally and behaviorally connected to their job and company. Another 16% of millennials are actively disengaged, meaning they are more or less out to do damage to their company. The majority of millennials (55%) are not engaged, leading all other generations in this category of worker engagement. Not engaging millennial workers is a big miss for organizations.[2]

A 2019 population estimate by the US Census Bureau shows that millennials are now the largest living adult population in the United States. That is significant to employers who must understand and engage with job seekers and current employees from this generation.

Gen Z

Generation Z grew up at a time when the Internet was always on and digital devices were at their fingertips before they could walk. They were preteens when the first iPhone debuted in 2007, and by the time they entered college or the workforce, using a mobile phone to access the Internet was the norm. They prefer to stream media, such as song albums or movies, over any other method. Video chats, instant messaging, and social media are normal ways they communicate with friends.

Interestingly, even though this generation was born digitally enabled, the preferred method of communication for Gen Zers is face-to-face. Generation Z is not afraid to learn new skills. Managers should consider letting this generation help lead the way with adopting new tech in the workplace and using their naturally collaborative talents to help acclimate the team to new systems.

This generation grew up during the Great Recession of 2007–2009 when the real estate bubble burst. During that period, there was a time of high unemployment and home foreclosures. Seeing parents and other adults around them worry and struggle had a definitive impact on this group. This could be why they have a focus on the security of benefits and competitive salaries when looking for job. According to research conducted by specialized staffing firm Robert Half, 77 percent of Gen Z believe they will need to work harder compared to those in past generations to have a satisfying and fulfilling professional life. This may also impact the fact that Gen Z has the hardest time maintaining a healthy work-life balance of all of the generations in the workforce. They are also natural multitaskers who are accustomed to operating multiple digital screens at once. I want to mention that at the time I am writing this book, the youngest Gen Zers are only 20 years old. They are somewhat newer to the workforce in comparison to the older generations. This means we do not have as much historical data on work habits over the life of their career.

Being accustomed to multitasking and having a poor work-life balance are the key reasons they had a higher rate of burnout than any other generation during the pandemic. Managers should keep a close eye on them and maintain a high level of engagement. Encourage down time and encourage them to not check work messages when they are off for the night.

Additional information from the research conducted by Robert Half shows that honesty and integrity are the top qualities Gen Zers would like in managers and employers. Opportunity for career growth was the most commonly cited career priority. Many would prefer to take on new roles and go higher up the ladder in the same company rather than look for new employment elsewhere.

Gen Zers are very practical by nature but at the same time are very ambitious. They have a high rate of entrepreneurship, and they are the most diverse of any previous generation, which helps them thrive in a diverse work environment.

They understand that the job market for them is highly competitive. They are eager to acquire hard skills to make them more valuable to employers. They would greatly benefit from mentoring programs and opportunities to learn soft skills in business.

Common Ground

When your company is pushed into a sudden remote work situation, you will not only manage the hurdles to the business that come with that change, but you must also help manage the fears and expectations of the team. By understanding their insecurities, priorities, and needs, you can help manage how they respond to the event and how well they adapt to the new reality.

As you learned in this chapter, they have different experiences when entering the workforce and different world events helped shaped how they see that world. But many of their needs are still similar. Remember that as the whole company is forced to go remote, people will be outside of their comfort zones and put into a place of uncertainty. As the saying goes, they may all be in the same storm, but they are not all in the same boat. People who may have never worked remotely will now be doing so full time. People who are newer and feel more dependent on the guidance of their peers and managers will be working alone.

To alleviate stress, increase authentic engagement, and enable a more productive WFH scenario, managers can emphasize those common needs when they choose how they engage employees as a team and as individuals.

Flexibility

A WFH atmosphere that allows for flexible hours will help everyone on the team regardless of their generation. Many boomers, Gen Xers, and millennials are enjoying life as independent contractors because of flexible work hours. Gen Zers also benefit from flex hours, along with encouragement from managers to maintain a better work-life balance. This gives everyone more time to spend with family and pursue other interests, in order to be healthier individuals.

The metaphor of an employee being chained to a desk all day is not really about the workplace but more about the huge blocks of time out of one's life that a person has no control over. It is the hallmark of the unhealthy and stressed-out workaholic. By getting rid of these shackles, you empower your employees to do more with their time. That type of freedom is invigorating, and you will find that your employees' zeal for their work and the quality of their work will go up.

Providing flexible hours means a shift in the way people are managed, but it shouldn't mean a shift in goals. The goals remain the same: retaining good people, keeping up productivity, increasing engagement, maintaining company culture, and staying on-mission. By allowing flex hours, you help cut out the trivialities of managing people and you will have a clearer ability to focus on what truly matters to the company.

Flexible hours are not time-oriented; instead, they are result-oriented, goal-oriented, and trust-oriented. To have a flexible work policy that works well, communication and expectations must be clear. Expectations should include clear direction surrounding due dates, availability, communication habits, responsiveness, and responsibilities. This helps workers manage their new levels of personal accountability. They will also have new levels of independence from supervision, which can cause anxiety in both employee and manager, at least at first. Stronger engagement efforts and consistent levels of feedback will help mitigate that anxiety.

Feedback and Positive Reinforcement

There are many avenues for engagement and communications with the remote team, but feedback is one of the most critical. The feedback we will discuss here is not the type of feedback that you offer at quarterly or annual reviews. Instead, it should be honest daily feedback that helps manage the workflow and employee satisfaction. At its core, it helps keep everyone on track, but it can do much more. Good feedback given consistently will reinforce a positive and healthy work environment, keep up morale, and maintain personal engagement.

Good feedback on a regular, consistent basis helps provide direction, not just for tasks, but for all aspects of remote work. It can start with the daily check-in to see how everyone is doing. Review some of the tasks from the previous day or from earlier in the week. Provide feedback on those, but if you have some feedback that won't be very positive about an individual, do that in a more private setting. It is said that it takes seven positives to reverse the effects of a single negative remark. Structure your feedback with that in mind, especially with younger or newer employees. Help them walk away with deeper knowledge and clearer understanding of expectations.

Group feedback is important because it helps the team know that they function well as a unit. It helps them feel supported, helps them know areas that they need to strengthen, and enables them to be more effective as a unit.

Consistent feedback creates more opportunities for coaching and mentoring. Leadership responsibilities should involve coaching and mentoring. With work-from-home situations, some of that coaching will be about helping workers balance out their roles as an employee and as an individual who is balancing a healthy at-home life as well. You can ease the uncertainty and anxiety your team will face by modeling effective ways of handling work issues and by reminding team members to take downtime. Reinforce that they are valuable members of your team and that it is important to stay mentally focused and healthy. In this new and sudden WFH, you will not just coach on how to do the job, but you must also be emotionally supportive. This is part of authentic engagement, which is supportive on a personal level as well as a professional level.

When it comes to feedback, the more specific the better. Go into details about where there was strong work done, how it affected the outcome of projects, where there is room for growth to meet end goals. Good feedback leads with appreciation and wraps up with end goals in mind.

Appreciation and Recognition

One of the top factors reported by people who suffered burnout during remote work periods of the pandemic was that many felt unrecognized for their hard work. Under normal circumstances, feeling unrecognized can shrink an employee's morale. In trying times and in uncertain conditions, it can devastate a person's self-esteem, their sense of belonging, and their sense of being valued.

One reason for burnout in remote workers is likely the lack of recognition an employee feels when they are already struggling. Part of recognition is awareness. Seeing and listening are big components of that awareness of your employees. If you see your team members looking haggard, unkempt, or stressed, recognize those red flags and engage the employee. Recognition for doing great work is important, but leaders need to fine-tune their skill sets for remote workers. Recognition of those who are struggling is an important part of appreciation as well. When you truly appreciate someone, you care for them. You want them to do well and be well. If you recognize signs of them not being well, make sure they know that you see the changes in them and that you want to help. It could be that something as simple as taking some paid time off or adjusting their workload will give them the ability to decompress and will make a huge difference for them. Knowing that you recognized that they were struggling and that you engaged with them so you could help makes them feel valued and increases a sense of belonging.

Show your teams and individuals that you appreciate them and their work. Do it publicly and do it often. Make announcements in which you thank specific employees for a strong work ethic or commitment to company values. Use group chat channels or other similar avenues for general company announcements. You could even applaud them on your company social media channels. Make your gratitude something that is expressed without restraint. A little appreciation goes a long way, but consistent recognition along with appreciation creates an environment where people feel understood and valued.

> **BEST PRACTICES**
>
> Regular feedback provides direction for all aspects of remote work. Feedback followed by recognition and appreciation helps team members feel valued and helps them understand that they are a vital part of the company.

Commitment

Just as an employer wants a commitment from those they hire, employees want to feel that their company is committed to them and their goals. Members from each generation will have different personal and professional goals, but your commitment to each should be equal and inclusive.

Engage them to see what each has in mind for personal and professional development that can help them with both their life and their careers. Many employees will look for the opportunities to learn new skill sets and expand their roles in the company. Some will appreciate the chance to be leaders and mentors, while others would relish the chance to be mentored in soft skills, business networking, and relationship building. Help match these desires with the strengths of others on your team. These complementary strengths and skill sets keep a cycle of both wisdom and experience-based knowledge flowing through your company culture.

By helping create development road maps, you increase the value of each team member while increasing engagement and overall job satisfaction. Those are elements of a company culture that people want to be a part of and learn from. But if they don't, that's okay. As Marla DiCarlo from Raincatcher points out in the following interview, not everyone may be cut out for your company. You need the right people who fit into your company culture and values. We dive into the details of company culture and how it affects remote working in the next chapter.

MARLA DiCarlo, CEO AND CO-OWNER RAINCATCHER, LLC

Company Profile

- Location: Denver, Colorado
- Employees: 20
- Primary Offering: Raincatcher, LLC, is a nationally trusted business brokerage company that helps entrepreneurs buy and sell remarkable businesses.
- Primary Audience: Small business owners with revenue between $1 million and $10 million.

About Us

Raincatcher guides entrepreneurs through the business buying and selling process in order to get the maximum value for their business. They are real entrepreneurs themselves who believe small business owners and their clients are the heartbeat of America and deserve their chance at the American dream. Raincatcher uses the latest in technological advancements and integrated digital marketing to connect the best buyers with their sellers. They provide value-building services as well as bringing merger and acquisition practices to their brokerage, transforming any size business into ready-to-sell enterprises.

What apps, services, or technology do you use to bridge the gap from in-person to remote in order to keep the workflow alive with your team?

We have found several apps to help our team bridge the gap from in-person to remote. We use Google Chat to bridge the gap with quick internal communication, such as a quick question you would swing by someone's office or make a quick call to ask or notifying the team about a win or something important. We also use Google Meet to manage our video and teleconferencing. We have set up several team-building meetings to help keep employees engaged. We do a monthly book club, a biweekly happy hour with a theme, and a biweekly all hands on deck, where we provide department updates and training/education on one topic. We also encourage our team to turn on their videos during these meetings. It helps to see each other and read body language.

We also started using an app called VideoAsk, which provides an effortless way to have asynchronous video conversations, which we use to engage our team, customers, and new leads. Each week a team member records a "challenge," where they ask a question and everyone on the team records a VideoAsk responding to the question. It is a great way to keep employees familiar and comfortable with recording videos, and we have found it helps our team to bond as everyone learns more about their team members.

We also use a platform called Lattice to help drive performance and to assist with employee engagement. This is a great platform to set and track employee, team, and organizational goals, such as progress moving toward our BHAG (Big Hairy Audacious Goal). It is also a great way to provide feedback and recognition to team members who have gone above and beyond. We use this feedback for quarterly bonuses and tracking employee progress toward our Raincatcher Drop award, which is an award given to one employee each year who demonstrates our core values, helps our team toward our goals, and helps the company as a whole.

You mentioned that you transitioned your previous company to go fully remote. Can you tell me a bit about that?

I owned an accounting/bookkeeping, fractional CFO business, Kaizen Business Results. I started the company with everyone coming into an office. The problem was we were growing quickly, and I started to run out of office space. Within two years of starting, I made the decision to go 100 percent remote. I had about 20 accountants, bookkeepers, controllers, and CFOs at that time, so there was a lot to consider. How could I ensure work got done on time, how would I track time, how would I train employees, how would I onboard new clients and answer employee questions as they started working on the new accounts?

The first thing we did is we found a great project management, time tracking, and billing software; we chose Zoho. This was new software back in 2012, but it was a great option since it billed per user and allowed us to add on features to our software as we grew. We set up templates to make it easier to onboard new customers. We created customer manuals, so it was easier to have new people work on an account if the regular employee was out of office. We created automated workflows to request documents and follow up with customers so our employees could focus on getting the work done instead of chasing after documents.

We also did a lot of team building and had weekly huddle calls to discuss accounts, training opportunities for the internal team, and all-hands-on-deck meetings so employees could discuss bottlenecks, wins, losses, and personal issues they might have going on at home.

What was the most difficult part of going fully remote for you and your team?

Reminding everyone that they had a team out there even though we were not together in an office. Employees would often become siloed and not reach out for help from their team members. They would try to problem-solve issues instead of reaching out to their team to get support. We put a policy in place that if an employee worked on an issue for more than 15 minutes, they would stop and reach out to their team to get help. Back then we were using Skype as our team chat platform. Another thing we did is weekly training opportunities for the entire team. We encouraged team members to attend live but if they couldn't we recorded the sessions. Each week a member on the team would present to the group a training topic that added value to working with our customers, such as best practice to create QuickBooks customized reports, how to reconcile books and fix common issues with reconciling, and how to handle conflict and diffuse a difficult situation.

After about one year, everyone on our team embraced working from home, and I found as an employer it helped me to find high-level candidates without having to pay them top salary because of the benefit of working from home.

What was the easiest part of going remote for you and your team?

Not having the commute back and forth to our office and being able to spend more time with our families. Once we started working from home, many employees commented on how they did not realize how stressed they had been with the drive in, missing their kid's events, being short with their spouses because they were running late, and feeling like a failure at home because they were not there to manage the day-to-day.

How was efficiency affected with your team being remote as opposed to working traditionally? Was the communication of ideas affected when done remotely?

The number one thing that comes to mind is getting everyone out of the mindset that they are working alone, in a silo, and forgetting they have a team. One of the benefits when we were in an office is when someone had a problem, they just went to another team member or knocked on an office door. So, we put the policy in place that they had to stop working on an issue after 15 minutes and reach out to their team for support. We also started referring to Skype calls and chats as "knocking on the office door"—if you were busy, just let that team member know when you would be free. We also worked on the mindset that there are no stupid questions. Raise your hand and ask a question if you are not sure about something. There was no reason to guess or spend time trying to solve problems; ask your team.

Did you create any avenues or methods for your people to stay social with each other?

Yes, this was the game changer. We sent birthday cards with a small gift card under $15. We also sent an email to the team, and their team members would send GIFs and best wishes. We also celebrated annual anniversaries, big wins, and special events with a card and email recognition. We did monthly happy hours and annual summits. The annual summit was the best team building.

With Raincatcher, we normally did two per year: one in the summer where spouses were invited and it was a "fun" get-together, and the other was in January, where employees attended and we did a review of the last year, planning for the upcoming year, and team vision and goal setting. This year because we could not get together, we did a virtual event. We did two days for six hours and provided Grubhub gift cards so everyone could order lunch and eat together. We did a Raincatcher Jeopardy in groups of four with a Grand Finale competition. We left time to do check-ins, brainstorming midday, and takeaway at the end of the day. It went really well. Everyone enjoyed the event, felt like they bonded as a team, and appreciated the leadership efforts to make everyone feel included.

Do you adapt your management styles when it comes to engagement with different generational team members in the remote environment?

Yes, I feel as a leader you have to adapt to everyone on the team in order to have loyalty and fulfilled team members. I have found that baby boomers and Gen X need structure and organization: simple things like providing an agenda for meetings, creating process manuals, organizational charts, and goals.

Millennials and Gen Z like collaboration. They want to be "heard" and feel like they are contributing to building the company. We put in place collaboration team meetings, opportunities to work on projects in other departments and submit new ideas on a form that is reviewed each year by leadership, and 360 feedback.

What are some of the values of your company culture that keep the remote environment working so well?

We live and breathe culture in our organization. Everyone knows how our mission and core values contribute to reaching our goals and achieving our BHAG. We discuss how culture equals action, meaning the tenets of our company are something we do every day, by everyone; it is how we walk the talk and how we are judged as an organization by our customers and our peers. I believe because we spend so much time discussing the importance of culture, this helps our team to be unified in a common vision, so even in a remote environment everyone knows what our company believes in and why we are the best.

We won "2020 Best Workplaces" by *Inc. Magazine* during COVID. I received the news that we won April 2020 right after we had temporarily reduced our employee hours. I remember getting the email after I had finished the last phone call with one of my employees, and I said "Yay, what do I do with this?" but we used the award as a reminder that though we are going through tough times, we are the BEST, and we will overcome and get back to being the BEST.

We brought everyone back on full time by the end of May and since then have added eight new people to our staff and continue to grow.

What would you do differently?

I would make sure I had everything in place from the beginning. I would encourage employee feedback so you can use it to improve and grow. I would encourage collaboration and celebrating success individually and among the T.E.A.M.: "Together Everyone Achieves More."

What advice would you give?

Make sure you hire the right people and you have the right people filling the seats on the bus. If you have someone that does not buy into your culture or the vision of the company, get rid of them. One person can destroy a team. Remember that you need to provide an environment where your people feel fulfilled and they are working on goals—if your people are coming to work just for the paycheck, you have a problem. Find out why they get up in the morning, what motivates them, what is their personal "why," and then do goal setting to help them get there.

Notes

1. https://www.purdueglobal.edu/education-partnerships/generational-workforce-differences-infographic/

2. https://www.gallup.com/workplace/231587/millennials-job-hopping-generation.aspx

Creating the Right Remote Team Culture

There are many articles and blogs written that discuss the importance of company culture in large corporations, but company culture should be considered an important factor for companies of every size, from a one-person shop to a small business and beyond. The culture is what drives the people within the company to make the decisions they make each day. It drives how they interact with customers and coworkers; it defines how they act or react to unforeseen circumstances. In a large part, company culture represents the essence of the company.

In this chapter, we discuss company culture, how to adjust it to meet the remote workers' needs, and how to maintain that culture when the team is distributed. If you take care of the company culture, everything else will take care of itself.

You Already Have a Company Culture

Every business has developed a company culture that guides the everyday actions of its people. However, it is not unusual that a company leader is not aware of their own company culture. When this happens, the culture is not being guided and shaped by leadership. It is the result of a lack of intent-based leadership. Intent-based leadership is about purposefully creating an environment where people feel valued and instilling in them that their work has a purpose that contributes to something bigger than the individual. When this type of leadership is absent in a company, it is reflected in the attitudes of the employees when they interact with customers and with each other. Further, not all companies are consciously aware of the elements that are defining their company culture. Regardless of whether they are aware or not, those elements are already impacting the organization. The aspects of company culture are not set by the words written on a memo or shared documents from human resources. They are defined by the actions of the leadership of the company and the rest of the team. For example, if leadership typically acts in an unempathetic way toward employees, then employees will follow suit with each other and clients. If leadership watches the clock and makes sure that people are punching in on time or rushing employees to punch out on time to avoid overtime pay, then intentionally or not, they will create a culture of clock watchers. Those employees will be stressed with each start and end of the day and won't bother to worry about the quality of the work they did in between. They work for the hours.

Company culture goes much deeper than what is discussed in just these examples, but the point is that it is important to create your company culture with intent-based leadership. Do not simply allow it to be created on its own and defined organically without the clear guidance of your vision for the company. If you did not create your company culture with intention based on your values and your mission, then you need to start now.

BEST PRACTICES

Create your company culture with intention based on your vision for the company and its values.

What Is Company Culture?

A corporate or company culture is exemplified by how its people act and interact in their day-to-day business. It comprises the shared values of the organization as well as shared goals. For example, some of the core values of American Express are as follows:

Integrity: We uphold the highest standards of integrity in all of our actions.

Teamwork: We work together, across boundaries, to meet the needs of our customers and to help our company win.

Respect for People: We value our people, encourage their development, and reward their performance.

Personal Accountability: We are personally accountable for delivering on our commitments.

Quality: We provide outstanding products and unsurpassed service that, together, deliver premium value to our customers.

From these elements, people within the culture understand right from wrong. These values let them know what behavior is acceptable and what is discouraged. When a company clearly defines what their core values are, what their company mission is along with their goals, they begin to define their company culture with intention.

When broken down to the most basic elements, the formula for a company culture is Values + Behavior = Culture.

The culture is greater than the company itself. The culture represents an ideal to strive toward. It is a marriage of shared values that inspire others to do better work in all aspects of their lives, not just their place of employment. It inspires commitment to realizing the ideal via their actions. Purpose-driven work is meaningful work.

Multiple studies show that both millennials and Gen Z desire purpose behind the work that they do at their jobs. The word *purpose* in this context may seem a bit ambiguous, and that is because it means something different to each person. However, the idea of having purpose can be quite simple. Gallup's report "How Millennials Want to Work and Live" defines purpose as "liking what you do each day and being motivated to achieve your goals." A strong company culture can have a powerful effect on productivity, but more importantly, it drives a better quality of work and deeper employee satisfaction.

Impact on Remote Work

Your company culture acts as a guide in what to do when nobody else is around. In other words, if an employee was faced with a decision but there was no supervisor to ask for advice on what to do to do, that employee would then act based on the company culture by default. If left with only option A or option B as solutions, the employee would have an understanding of which option is the correct choice based on what they understand of their culture.

Because of this, the cultural understanding is of utmost importance with remote employees who are not under the watchful eye of a supervisor and who may have flexible hours. Remote employees will have a deep understanding of what is expected of them and how to respond in nearly any work situation within an organization that has a strong company culture.

In Times of Stress

The company culture is the best tool to help your people adapt and thrive in times of adversity. The culture of your company is ever present, regardless of whether the person is in the office, working from home, or on the road. Your company culture helps shape the behaviors and attitudes of each person working in the company. Their attitude helps set their outlook on events and how they handle events that may push them out of their comfort zones.

The company culture is also a source of strength. It builds confidence. People know what to do in nearly all situations based on their culture. A well-engaged remote employee would know how to act, how to communicate, and what their company expects based on what they know of the company values and mission that are exemplified through the company culture. The well-engaged remote employee will be able to discern what to do even without any supervision or peers to turn to for guidance. Their ability to rely on and trust that knowledge fuels their confidence at work.

The company culture is a huge contributor to the various goals of the company from sales to support to marketing and the hiring process. In fact, it is a main component to the success of the company.

Company Culture and Success

> Maintaining an effective culture is so important that it, in fact, trumps even strategy.
>
> *—Howard Stevenson, Professor Emeritus,*
> *Harvard Business School; Director of Publishing*
> *and Board Chair of the Harvard Business Publishing Company*

That is a powerful statement, and it's not an exaggeration. In a healthy company culture, each employee feels like part of the larger team. As a team member who wants to belong, they want to see the team succeed in its goals. They have an attachment to good outcomes for the team and are willing to put in the effort to achieve those outcomes. They work toward that greater purpose. Purpose-driven work is also high-quality work, and that type of work reflects well on the whole company, helping to drive its success. That drive toward excellence is reflected in the work across all departments. It would be a touch point with every customer, each new hire, and all existing employees.

When a company emphasizes its core values in everything it does, its culture drives success at every level. For this to work, it is critical that the values of the company align with the values of the people it hires. That does not mean that everyone must have the same set of beliefs, but they all must share what they hold as important when it comes to core values. It is your personal values that guide you in your decisions and motivate your actions. I discuss in detail how aligned values and company culture impact company success in my book *The Artful Ask, Pimbleberry Publishing, 2019*. Here is a snippet from that section:

Understanding that employees tend to enjoy their work more when their personal values align with the company's value system should never be forgotten when cultivating company culture. They will create better relationships with other team members and are far more productive. In fact, a study of corporate culture and performance revealed a staggering correlation between the two.

Two Harvard Business School professors did a long-term study of 200 major corporations. The results are published in their book *Corporate Culture and Performance*, which shows how culture has a definite impact on the financial performance of a company.

This chart shows the results over an 11-year period of 12 companies that had corporate cultures which valued employees, owners, and customers. These company cultures also encouraged leadership from everyone on the team. The study compared these 12 companies to 20 companies that did not have such a culture. The financial difference is staggering.

AVERAGE INCREASE IN . . .	20 COMPANIES WITHOUT PERFORMANCE-ENHANCING CULTURES	12 COMPANIES WITH PERFORMANCE-ENHANCING CULTURES
Revenue Growth	166%	682%
Net Income Growth	1%	756%
Stock Price Growth	74%	901%
Employment Growth	36%	282%

Source: John Kotter, James I. Hesket "Corporate Culture and Performance"

> As you can see from the graphic, company culture is wildly impor-
> tant to a company's profitability. It is also a driving factor in
> attracting and retaining great talent.

That last line is an important one because getting the talented individuals that fit into your company culture starts before the hire. A company
should position itself to attract the best talent and best team members
that it can. That means having a company culture that is attractive. If you
have an unappealing company culture, it is reflected in how employees
feel about the company, and the outside world will know about it as
well. In this day of social media and Yelp reviews, it's easy to find out
what it's like to work at any given company.

BEST PRACTICES

Companies should emphasize their core values in their actions, behaviors, and
communications with everyone they interact with—from employees to vendors to customers.

Retention: The Cost of Replacing Employees

You can understand the importance of a good company culture by looking
at examples of companies with negative company cultures. We can imagine
a scene from a movie in which a customer walks into an empty super-
market. The camera pans to a cashier who is leaning against the counter,
chewing gum and lazily scanning items while avoiding any unnecessary
interaction with the customer. Somewhere in the background, there is
a stockperson with earbuds in putting cans on a shelf in a clumsy and
careless manner while a manager reads a tabloid from behind a Plexiglas
booth positioned for observing the workers. Although a bit cliché, it's
a great example of a bad work culture. I'm sad to say that I've walked
into plenty of businesses just like the one I described.

Needless to say, that sort of company culture does not cultivate a
great customer experience or brand experience, and it's even worse as
far as the employee experience is concerned. The customer at least gets
to leave the place after a few minutes of walking through the doors.
You have to ask yourself who would want to work in an environment
like that, or for a company that does not consider their employees to
be a team. Instead of creating a team-based environment, it becomes an
hourly job with a loosely associated group of people who don't seem to
want to be there at all. It's a culture that is entirely transactional, where

someone is paid to show up, to accomplish a set number of tasks for a set number of hours, then to leave, only to repeat the process during their next shift. With this type of company culture, you might wonder who would want to stay working there for very long.

We've glimpsed at what a bad company culture is like for the customer and the employee. Now we will consider whether a mediocre company culture is much better. When it comes to retention, I think the answer is no. It's one thing to leave a miserable job to get *any* job that's even marginally better, but if the new one is still not great, you won't be looking for reasons to stay.

A healthy and attractive company culture is not just a great recruitment tool, it is an excellent employee retention tool. A 2018 study performed by employee experience platform provider WeSpire showed that 75 percent of Generation Z thinks that work should have greater meaning beyond the paycheck. They take the time to read the mission statement and understand the company values when job searching. The study shows that Gen Z is the first generation to prioritize purpose over salary when it comes to their work. At the same time, the study showed that Gen Zers prioritize authenticity and prefer to be able to respect the managers that they work under. The study also showed that Gen Zers will be happy to actively publicize ugly corporate cultures or toxic work environments. This means that transparency is big with this generation of the workforce and they are happy to leave companies that have an unhealthy or even unsatisfying culture, even if the next job is for lower pay.

Figure 10.1 summarizes statistics based on a 2018 annual survey by recruitment site FlexJobs, a 2018 employee retention report by employee engagement company TINYpulse, and a 2018 study performed by employee experience platform provider WeSpire. By looking at these statistics, you can see what is important to workers and how these priorities would be reflected in a company culture.

Controllable Quit Rate

According to a 2019 study by employee retention and engagement company Work Institute, it costs the average company $15,000 to lose a US worker. From trends reported by the US Department of Labor, the Work Institute study estimates that one in three US workers will voluntarily leave their jobs each year by 2023 for better opportunities. This shows a steady trend, considering that in 2018, 27 out of every 100 US employees quit their jobs. When looked at collectively across all US employers, this high turnover rate equates to roughly $1 trillion

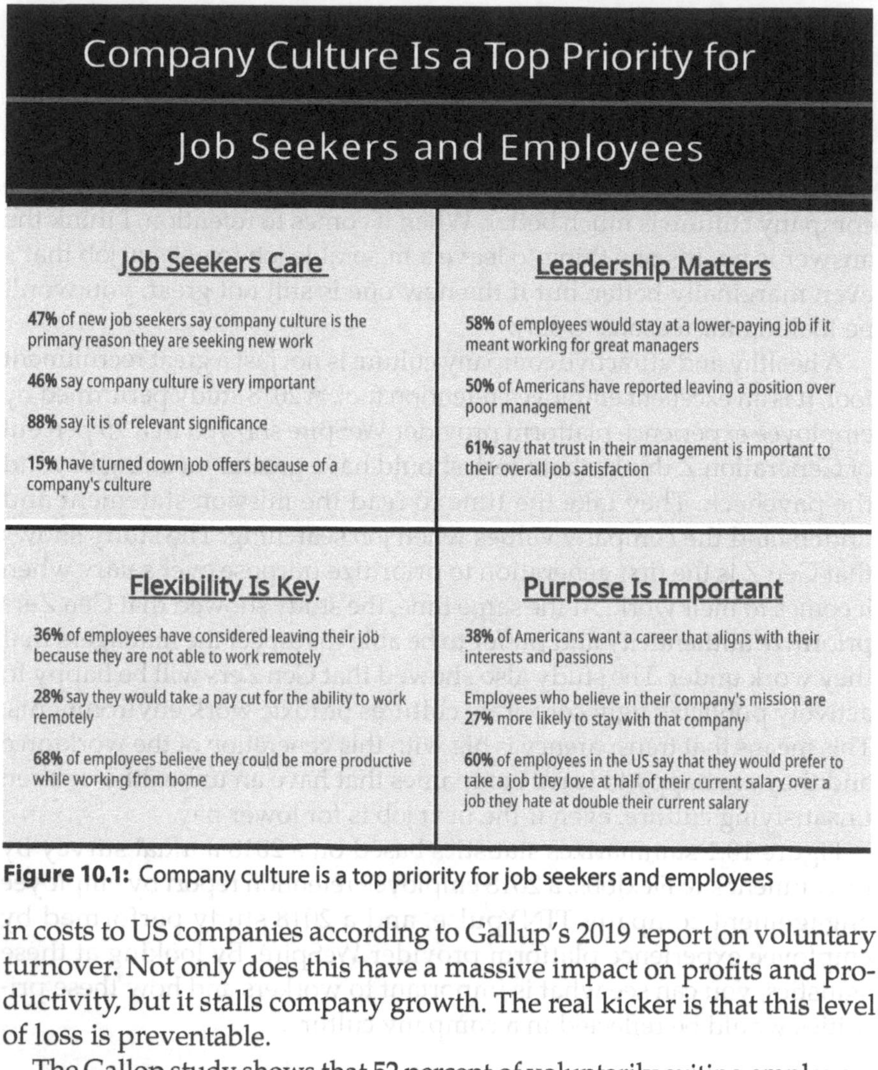

Company Culture Is a Top Priority for

Job Seekers and Employees

Job Seekers Care

47% of new job seekers say company culture is the primary reason they are seeking new work

46% say company culture is very important

88% say it is of relevant significance

15% have turned down job offers because of a company's culture

Leadership Matters

58% of employees would stay at a lower-paying job if it meant working for great managers

50% of Americans have reported leaving a position over poor management

61% say that trust in their management is important to their overall job satisfaction

Flexibility Is Key

36% of employees have considered leaving their job because they are not able to work remotely

28% say they would take a pay cut for the ability to work remotely

68% of employees believe they could be more productive while working from home

Purpose Is Important

38% of Americans want a career that aligns with their interests and passions

Employees who believe in their company's mission are **27%** more likely to stay with that company

60% of employees in the US say that they would prefer to take a job they love at half of their current salary over a job they hate at double their current salary

Figure 10.1: Company culture is a top priority for job seekers and employees

in costs to US companies according to Gallup's 2019 report on voluntary turnover. Not only does this have a massive impact on profits and productivity, but it stalls company growth. The real kicker is that this level of loss is preventable.

The Gallop study shows that 52 percent of voluntarily exiting employees say their manager or organization could have done something to prevent them from leaving their jobs. Fifty-one percent said that in the three months before they left, neither their manager nor any other company leader spoke with them about their job satisfaction or future with the organization. That is an incredible lack of employee engagement.

Companies need to make a conscious effort to attract the best talent and then retain that talent once they are a part of the team. They must mitigate the financial costs and talent loss that comes from employee turnover.

A strong and healthy company culture ensures that the quit rate is far more controllable. As has been shown with Generation Z and millennials, more financial compensation is not enough to make the job more satisfying. Employers and company leaders must leverage the power of their culture to shape the company into something that people want to be a part of and help nurture. That culture may begin in the physical workplace, but it also has to be present in the distributed teams who are working remotely.

BEST PRACTICES

Strong engagement and a healthy company culture should be priorities for company leaders to ensure employee satisfaction and lower the turnover rate.

Shape Your Culture with Remote Work in Mind

Take a deep look at your current culture to see if it is where it needs to be to best serve your remote teams. Are the ways that your company values are expressed in your policies still relevant when it comes to remote workers?

For example, let's take a look at communications policies. As Leslie Murphy from Raybourn Group International stated, since the COVID-19 shutdowns, their policy for videoconferencing has changed when in-office and remote employees are on a staff meeting together. That new policy is staying in place for good. Everyone in the office will now participate in the videoconference from their own desk, as opposed to having the in-office employees together in the conference room while the remote team members call in from their individual spaces. This communication policy looks at the company culture for its direction. It's a choice to eliminate any perception of there being two distinct groups of employees: those who work in the office and those who are on the outside looking in. With this new communication policy, everyone is now on the same level as a single, unified team, even in how they communicate.

This policy is in sync with company values that emphasize the importance of its people and cares for their well-being by taking intentional steps to decrease feelings of isolation. This type of intentional tweaking of policies and methodologies must take place to ensure that the remote team members are considered to be full members of the group who are equal in every way to their peers inside the office. It's not enough to simply believe in this ideal; company leaders must also show it via their actions and decisions.

Leaders Set the Tone

The leaders of the company are the ones who have oversight of the company and who look at the big picture. Taking a look at how company leaders communicate with employees and how employees communicate with each other is just part of the big picture.

Those in any form of supervisory position must understand that they take up the mantle of stewardship for the company culture. How they act, and the tone and manner in which they speak to those they are leading as well as those above them in the company hierarchy, reflect upon the company and its culture. Transparency, honesty, and openness are important in a company culture but are critical when it comes to remote workers.

In order to minimize feeling siloed or out of the loop, remote teams depend on openness and transparency. It must be consistent at every level of interaction. That brings authenticity to those company values. It empowers those aspects of the company culture. It cannot be faked or done just with some of the people in the organization. That transparency must be ubiquitous and must be ever present in the actions of the company leaders.

Do as I say not as I do is never part of a healthy corporate culture. When that happens, there is a loss of transparency and a lack of clear guidelines for behavior. All individuals must be equally accountable to the company values and mission if the culture is to remain strong and well defined. Behavior is critical to the formula and that means the behavior of all, but especially of those in a leadership position.

Workplace Transparency

For a remote team to be highly successful, transparency must be part of the company culture. We should define what transparency means with the distributed team and how to achieve it in the remote workplace. Workplace transparency begins with fostering openness in the company culture.

Informational transparency has roots in the accessibility of data and projects from departments, but there needs to be accessibility to leadership. Within a culture of open communication, workers should be able to ask any question to leadership of any level. A number of companies have created channels or some type of electronic forum for employees to engage with leadership in open, straightforward communications.

The move toward openness in the workplace for some companies has led them to be transparent with company financials. A good example of such openness and transparency is from the interview with Scott Baradell from Idea Grove.

"We don't believe in activity tracking. We prefer to lead with trust. If you don't trust your team to get the job done, whether at the office or outside of the office, why would you hire those team members? In addition, we have a set of clients that would let us know if we were letting them down, so we really have an immediate feedback loop letting us know if we have a problem. Finally, we are implementing Open Book Management using the Great Game of Business methodology, where all of the employees understand the financials and have a stake in the outcome of the performance of the company. This level of transparency helps keep everyone accountable to everyone else on the team."

Another aspect of transparency that facilitates openness is clarity in expectations and goals. This clarity should be on both sides of the employer-employee relationship. The employee will have career goals and have certain expectations of the company to reach those goals. The employer will have goals for the company and has expectations from those they hire to reach those goals. When everyone is open and honest about what they want, there can be real, straightforward, and honest communication.

If we take all these factors into consideration, I would say that the formula for transparency in the remote work environment would be:

Openness +Accessibility + Clarity =Transparency

Be Worthy of Their Trust

As shown in the WeSpire study mentioned earlier, Gen Zers prioritize authenticity and want to have respect for the managers that they work under. As a company leader, you are holding their careers and livelihoods in your hands. You are shaping aspects of their future. Take that responsibility seriously. If someone who worked under you were to leave the company to go work for another, what would they take away from their experience working with you? Will they be better for it? Will they be more fulfilled, or will they walk away thinking that they are glad to be out of there and away from you?

Although it is true of those in every generation, Gen Zers and millennials especially are skeptical and often do not take words at face value. They are guarded against being taken advantage of by employers and are pessimistic against large institutions.

If you are a leader or if you are training to be a leader, you must be authentic in all that you do. Be true to your word, and be honest about the intentions behind your actions and your decisions. You must be credible. You must believe in the shared values that you promote and only then will it be reflected with authenticity in your work and interactions. Only then will your engagement be truly authentic.

Define Your Vision Well

Your company mission statement is your North Star in that it helps to navigate where the company is heading. It helps you not only to reach business goals but to identify what those tangible goals should be. A vision statement is a tool companies use to help demonstrate the principles and values behind the work they do. It represents the aspirations of how the company will impact those it touches. To clarify, here are some examples of both vision and mission statements from well-known companies.

Company: Tesla

Mission: To accelerate the world's transition to sustainable energy.

Vision: To create the most compelling car company of the 21st century by driving the world's transition to electric vehicles.

Company: IKEA

Mission: Offer a wide range of well-designed, functional home furnishing products at prices so low that as many people as possible will be able to afford them.

Vision: To create a better everyday life for the many people.

Company: TED

Mission: Spread ideas.

Vision: We believe passionately in the power of ideas to change attitudes, lives and, ultimately, the world.

As you can see from these examples, the vision statement portrays the emotional attachment to what the company does. It helps show what the company values in its work. While both statements are goal-oriented, the vision statement represents the heart of the company, whereas the mission statement represents the brain of the company. One deals with the intangible; the other deals with the tangible.

It's a good idea to revisit the company vision every few years as well as in times of big change for the company. Consider how the vision and its values are reflected in your company culture. Clarify it, and communicate it well through your decisions and actions. Emphasize it when giving accolades and recognition.

At the end of the day, the vision defines the purpose of the work of the company, and purpose-driven work is meaningful.

Emphasize Meaningful Work

Meaningful work is important for the morale of remote employees, and managers should encourage their employees to engage in meaningful work over getting wrapped up in the number of hours they put in during the day. Prioritize the purpose of the work when you communicate with the team. A sense of purpose drives the passion behind good work and ties it to the emotional connection that leads to extraordinary outcomes.

The work your employees do should also have meaning to their career paths. The good work they do should positively impact their futures. One way to help your employees advance is to provide mentorship opportunities alongside your leadership. You can help people with their soft skills, groom individuals for greater roles in the company, create opportunities for growth, help them with their professional development, and engage them in conversations about their career paths. By doing so, you help develop future leaders and help ensure the work that they do today has more meaning and a higher purpose.

Stay Engaged

When you stay engaged with your people, they will remain engaged as well. Engagement is driven by the leaders of the company. When top leaders remain engaged with their management teams, the managers tend to stay engaged with the teams under them. Engagement at every level is key. I encourage it as well with subcontractors, independent contractors, and vendors. Having a purely transactional relationship with contractors or vendors does not serve the company culture well. It is not inviting nor is it a good brand experience. To be trusted, a company culture must be authentic in its engagement with everyone, which develops into credible, consistent leadership that employees can believe in and respect.

Get Feedback and Act on It

You must be willing to make changes to processes and policies as needed, especially when you are faced with an abrupt change to the company, such as going fully remote. There will be things that simply don't work in a remote setting or with a hybrid team. Flexibility is important to the remote work culture, and this means management has to be flexible too.

It's not enough to be open to feedback, you must actively seek it out. Solicit feedback from your teams. Find out what is working and what is not. After a project is completed, ask the team what should or could be done differently the next time in order to improve the process for everyone. You will discover new ways that can boost productivity and increase overall job satisfaction.

Consistent feedback is necessary, but feedback without action is meaningless. Don't be so married to any process that you are unwilling to alter it. And don't be lackadaisical in acting on feedback by putting it on the back burner. That negates the purpose of engagement. When you get feedback, act on it. Acting on it quickly also cements the feeling of belonging for individuals who see their feedback acted upon. They can then see that not only are they valued, they are making a difference.

BEST PRACTICES

Staying engaged and acting on employee feedback will improve job satisfaction and help team members feel valued.

Be Clear on Digital Work Policies

Remote workers tend to have far less supervision, and flexible hours can mean different things to different people. To be both consistent and authentic, it is important to make expectations clear. Post your policies where everyone can see them at any time. It may be a single-page document, or it may be a folder with multiple documents on a shared drive. What's important is that the language of the policies is clear. They should cover every aspect of the digital work week.

When the COVID-19 shutdowns took place, larger organizations such as Cornell University had entire sections of their website dedicated just to remote working during the shutdowns. The main section included subsections with documentation for remote work, detailing expectations in areas such as management of remote teams, guidance on self-care and mental health, and home office needs.

Be as explicit as you need to be. Don't leave much room for personal interpretation. When everyone is on board with what to do, you get better work with less of a need for supervision.

Check Your Headings and Adjust Course as Needed

A company culture is like a living thing that moves and changes. You must constantly course-correct to make sure you are navigating the company to where you intended it to go. It's far too easy to get set off course if you are not staying on top of things. Unfortunately, it's also easy for bad habits or bad feelings to get ingrained in a company culture. Once in, they will spread to each new employee. That will undermine your efforts and the vision for the company.

Your company culture needs constant stewardship. Once it is where you want it to be, you must then maintain it and help it stay healthy. The culture should grow and expand to meet new needs even as the core values remain the same.

Your employees are a huge part of this stewardship. Make sure they know the importance of the role they play in it. Listen to them. Never assume that you know what they need or want. Be a facilitator in helping to adopt the change they need. When they are involved in shaping the culture, they are better engaged with it. As your company grows and some team members move on to new companies, others will come in and the culture will change to continue to meet the needs of its people and the company.

This is also why you must maintain and nurture your culture consistently but even more so when your company goes remote. The needs will be different, and the team must adapt to these new needs in order for the company and its culture to continue to thrive. A strong culture with clear expectations will help the entire team adapt more easily. It will also keep them better connected to the company while they are out of the office and physically separated from their team.

Maintaining Culture When Remote

Keeping up a strong company culture has proven difficult for many companies who found themselves suddenly going remote. The normal strategies they had used simply don't work when everyone is not in person. The strain of isolation and adaptation has caused many organizations to realize that, when the team is distributed, new methodologies and adaptations must be put into place to maintain the culture that keeps the company succeeding.

Be People-Centric and Values-Centric

All of the actions and communications from leadership must reflect the company culture when working with the remote team. During a time of high stress and uncertainty, going back to the core values of the company culture and emphasizing these values will help your people to function better. They will be struggling with being newly isolated from their team and with trying to balance their home and work lives. They will have anxiety about when things will go back to normal, or they may need to come to terms with their new normal if it may be permanent. The values of the company culture will provide them with a sense of normalcy and give them a familiarity that will act like a guiding light for them to follow. It is a strong foundation that will help steady the company in times of uncertainty. This counteracts the negatives that come from remote isolation.

When they see those values consistently reflected in the actions and words of their leaders, they will be well poised to adapt and continue to maintain their sense of being a part of the greater whole.

Get the Tech You Need

The technology you acquire should include the right collaborative tools, the right communications tools, and the right home office tools. It is important to invest in these because they will affect not only productivity, but the overall culture as well.

Having tech that does not fully serve the employees' needs impacts morale, job satisfaction, and increases the possibility of burnout. You cannot say that you are putting people first if you expect them to work with systems that do not meet their needs when working remotely. Doing so is neither consistent nor authentic. By investing in changes in tech to meet new needs, you show that you value your team and will rise to meet their needs as those needs change. As a result, you maintain the foundations of trust your culture embodies.

Ownership and Accountability

At many levels the remote team members are managing themselves. For that to work well, the foundation of trust must be strong, but there also has to be ownership and accountability for the work being done and its impact on the greater goals of the company. You want the team to think like stakeholders in the organization because they are stakeholders already.

Managers and leaders of the company create this sense of ownership via the company culture. It begins when new employees join the team, and it continues as managers communicate and interact with team members. What a stakeholder brings to the company is highly valuable, and the individuals should be treated in a manner that reflects the importance of their contributions.

Everyone on the team should understand how their actions and inaction affect aspects of the company, from sales to product development to client satisfaction. Everything they do impacts the brand image and the overall quality of the company. Team members need a deeper understanding of how these factors are all connected. They need to think and act like a vital part of the company and also own the quality of their work. Great work has a positive impact on the company. It resonates through all the departments, and it helps others to want to take ownership of their work as well.

To further facilitate the idea that everyone is a stakeholder who impacts the company, there must be accountability. Measuring productivity via goal setting and benchmarks and then the progress on those directives helps people understand the commitment involved with being a stakeholder. Employees should report consistently on the status of their assigned benchmarks. Then managers can see where they are and if they are on schedule to hit their mark and then adjust accordingly if they are ahead of or behind schedule.

This type of reporting and reflection must happen consistently in remote work so that team members are clear on what their goals are and feel confident in their direction and commitment to those goals. It creates a form of accountability that is not only about the individuals but is about a higher purpose of serving the group and the greater good. It sets the remote team up for success in self-management and in being a valuable asset to the group. It also helps them to think like true stakeholders in the company by showing that their work is impactful to the larger company goals and is purpose driven.

Employee Recognition

It is vital for remote employees to feel recognized for their work and the changes that they have to make to their home life and their professional life when the company goes remote. Being suddenly separated from their peers is something that they endure in order to remain with the company. Continuing to do quality work needs to be recognized, especially during difficult times. One of the technology segments that has had serious growth to meet this need is employee recognition software.

Employee recognition software helps companies meet the values of their company culture and maintain that culture with distributed teams by rewarding, motivating, and recognizing quality work and commitment. These platforms help strengthen connections among team members, increase employee satisfaction, and speak to the core values of the company by emphasizing what is important in the culture.

Bonusly

Bonusly is an employee engagement tool that helps to enhance company culture through recognition and rewards, which in turn improves employee retention and productivity. This system highlights and promotes company values by using hashtags for specific values during team-wide recognition posts. You can create hashtags for company values or team-oriented ones such as #teamwork or #customer-care. Then in the chats, when someone wants to give a team member some recognition, they can use the hashtags to add extra emphasis to the post. The use of emojis and hashtags give a familiar social media feel to the system and helps to encourage use. The system helps align company values with purpose-driven work through peer-to-peer recognition that is highly visible to all team members.

It offers a points-based system in which peers or managers can give points to an employee for doing good work. For example, say an employee got her part of a project done well ahead of schedule and that helped another team member to get their portion done faster. The second employee could go on the company Bonusly system and announce, "10 points to Carol for productivity and speediness!" in a Harry Potter-esque manner. The points can be given for practically anything and works well when used in the spirit of recognizing the effort and hard work of those on the team. Those points can then be traded in for rewards such as swag, donations to causes, or tasty treats to be delivered by online merchants. The system integrates well with a good number of collaboration and communication platforms that companies may already be using. The system tracks engagement and performance at individual and team levels, offering company leaders analytics on how employee engagement is going. These analytics allow for managers to catch sight of individuals who may seem actively disengaged with their work, which then lets those managers take active steps toward retention and increasing employee satisfaction.

Nectar

This system is designed to help maintain company culture, boost morale, and promote core values via social recognition and rewards. It is incentivized performance that highlights company values by showing meaningful recognition. Recognition can come in the form of public posts and announcements that are further boosted by rewards. Nectar uses a points-based system, and there are built-in contest options to increase employee accumulation of those points.

One feature that offers a great deal of opportunity for employee recognition, development, and retention is the Challenges feature. Challenges are activities that company leaders can create and customize to meet company goals, such as promoting employee wellness, learning new skills, or building social connections. These Challenges can be based on events or tasks, such as participating in mentoring sessions, walking 10,000 steps in a day, shadowing someone in a different department to learn new skills, or taking a course. It could also be something more social such as posting pictures of pets or taking a new hire to lunch. Points can be redeemed for gift cards, company swag, or something unique to the organization, such as a one-on-one lunch with the head of the company. It integrates with other popular platforms such as Slack, Microsoft Teams, and Google Chrome.

Fond

Fond harnesses the power of social recognition to help companies acknowledge accomplishments and milestones. Employees can redeem points for perks and rewards. These include items from major retailers as well as discounts from popular national brands and local businesses. The catalog of brands to choose from is truly massive. There are even options available for employee service awards, plaques, and trophies.

Managers can use Fond Perks to get discounts from outside companies and then use the perks to reward individuals or whole teams. For example, the managers could purchase movie tickets, concert seats, or store vouchers. A pair of concert tickets allows the employee to bring a friend or spouse and enjoy a night out on the town.

With Fond Rewards, managers or peers can send points to employees to recognize their hard work and efforts. As those points build up, individuals can redeem them for company swag, gift cards, or even donations to their favorite charities.

The system is easy to use and highly customizable. Analytics and reporting can be seen to help gauge employee adoption of the system as well as points redemption. The system gives the user the familiar feel of a social media platform to help encourage use. This allows for recognition to be seen by others in the company, which is an effective way to buoy self-esteem of employees and to help them to feel valued. It simplifies increasing the visibility of recognizing employees for doing good work and being strong team members.

Kudos

This is a recognition software platform that drives engagement, performance, and company culture. The system encourages recognition to help reinforce company values and to leverage that recognition to help drive employee performance. Recognition can come from managers or co-workers in the form of typed messages, awards, video messages, or rewards.

Kudos also emphasizes continuous real-time feedback. This is feedback from every level in the company, so employees can hear from managers as well as co-workers. It is also designed to help increase professional development, engagement, and overall employee satisfaction by highlighting a culture that values input, communication, and authenticity.

The system empowers social sharing by encapsulating communications into a digital hub. Birthdays, newsletters, photo galleries, employee profiles, announcements, and status posts can all be shown on the single platform. The system also has a feature called Spaces that acts as an internal communications space to post items such as wellness campaigns, community volunteering opportunities, and company best practices.

Cooleaf

Cooleaf helps to empower engagement, gratitude, and company culture by rewarding behaviors that exemplify an understanding of company values and expectations. Engagement comes in the form of social and highly visible recognition on the platform, such as awards, challenges, and events. The system uses machine learning and real-time feedback to help company leaders look at goals to see if changes must be made in order to reach them.

The system takes regular surveys and listens to feedback and input. It is then able to make suggestions on actions to take such as recognitions, professional development challenges, and others. This helps company leaders to better interface with remote employees and gain deeper insights into their needs and overall job satisfaction. This form of digital engagement assists in boosting morale by emphasizing that employees are valued and appreciated. Peer-to-peer recognition is encouraged by the social aspect of the platform. The use of hashtags in public and private posts helps give added appreciation and recognition for good work that aligns with company values.

Cultural Fit

As said previously, it is up to the leaders of the company to make sure that their mission and vision statements are well defined to help create the right company culture that exemplifies the values that are important to the organization. Once that is set, then you must hire the people who will be a good cultural fit. Bringing in a person that does not mesh with the vision and values of the company can be toxic. One bad hire can poison the employee bonds and trust that have taken a team years to build.

The following interview with Jeb Banner, CEO and founder of Boardable, touches on the importance of that sentiment. As he says, "If we make a bad hire, it is clear, and we hire slow, fire fast." The hiring process is important to get the right fit for your culture. That can be a bit different when hiring remotely. In the next chapter, we discuss the hiring process as well as other interoffice processes that can help keep your company thriving while remote.

JEB BANNER, CEO AND FOUNDER BOARDABLE

Company Profile

- Location: Indianapolis, Indiana
- Employees: 35
- Primary Offering: Centralization of all board meetings, documents, and activities in one easy-to-use place.
- Primary Audience: Small business owners with revenue between $1 million and $10 million.

About us

Boardable provides board management software for nonprofit organizations around the world.

What tech services/software did you use to go remote?

Our remote stack is Slack, Monday.com, and Pragli. We use Pragli as an always-on meeting space. We have offices in Pragli, so I can go into my (virtual) office and turn on my video, but lots of times you just leave your avatar on. Generally speaking, this is for huddles of 2 to 10 people, not more. There are different offices, so I can see who is active and I can request them. So this is our more ad hoc way of working.

We are currently using Zoom for our sales meetings and big team meetings. We have started to move over to Boardable video for this, since we have video-conferencing in the product now. Monday.com is for the project management, and then we have email as you would expect. So we are using Boardable meetings with video more for our own internal meetings, like leadership meetings, on-board meetings, things like that.

Do you do interviews and new hires remotely as well?

Yes, we're starting to hire much more, even internationally a little bit, and we just recently closed our series A of financing remotely.

I still have not met our investors in person, and we have probably six new employees that I have not met in person, which is pretty wild.

What is your remote hiring process like?

It's a lot of Zoom and Google Meet conversations. Today I have five interviews with different people, and it's quite the process. But, in a lot of ways it's more efficient, because you don't have as much transition time between things, and you don't have to worry about driving to meet somebody for coffee or whatever it might be because it's all being done from our homes.

How did working remotely affect you and your team working together?

I can't say that I necessarily prefer it. There are things about it I do prefer, but overall, I'd say that I'm definitely more of an in person leader. I like to be in the room, and one of the things we're struggling with is the osmosis side of running a business. When you sit around tables in a room, you're overhearing a lot of what's going on in other areas of the business. Then you can connect dots to what you're doing in your area of the business. We just don't have that right now. That's a big gap. So there's disconnects happening. Also, just working through problems; it's so much easier to work through something face-to-face when you have something going wrong. When you're physically present with somebody, there's the body language, the energy of the room, a whiteboard, and sticky notes. It's just a different dynamic.

Being in front of a screen all day versus being in front of human beings is a big difference.

How did you combat the feeling of isolation on your company?

I've become more engaged as a CEO, just making sure I'm having one-on-ones with people. I'm encouraging teams to have social times together virtually.

Happy hours, coffee, coffee breaks, there has been a lot of random gift giving. We did a birthday drive-by for somebody at one point; we just got everybody to drive by and do a little birthday thing. It was really cool. We really want people to feel like they're a part of a team, and that's super hard to do right now, especially when they haven't met anybody in person. So we're constantly looking to bring people into a non-business experience that is just a social hangout.

I will say this: I have never given away so many DoorDash gift cards. That's been a really nice way to just sort of say, "I know we can't have lunch together, but here, dinner is on us tonight."

How did this affect your company culture?

We are really fortunate to have this initial core group of people that have been with the business since very early on that I feel I can trust to represent the culture and to bring that culture into this next generation of hires. We're going to make some mistakes. We're going to make some bad hires. It's inevitable. My goal is to make sure that we've got a culture that is self-healing. If we make a bad hire, it is clear, and we hire slow, fire fast. I definitely subscribe to that. The culture recognizes that we're doing the things that are right in order to keep the team performing at a high level, and also filled with good people, kind people who care about each other. A caring culture is important.

We also don't want to build a monoculture. We've been fortunate to really move away from the sort of white male tech DNA that so often tech companies get stuck in. We have a female majority company, which is very unusual for tech, and an increasingly diverse team, which is very exciting.

What advice would you give?

I would say overcommunicating, job one, and overcommunication could mean daily stand-ups to start off, just to keep that team momentum going, even if it's just asking, "How are your kids doing?" Just be a little bit more present. Check in with people on a personal level, text them, call them, and ask them how they are doing. Because this is stressful, this is hard, and that personal check-in is really powerful, especially coming from leadership, so encourage your leadership to do the same.

On a more practical level, give them budgets to improve their home office. We did that, and I continue to push for it. Do you need a better chair? Do you need a desk? Do you need a light? Do you need a camera? What do you need? You're going to be working from home; how can you be comfortable in your house? You may have kids running around in the background. I get that. I'm not going to fault you for that. But how can you have as professional an environment as possible, to do your work in a way that makes you feel comfortable at home?

I think the third thing would be getting more into the tools you're using to run the business. Doing a full audit of that, and really assessing whether you need to bring something like Pragli or Slack into your business, where before it wasn't as needed. Looking at your toolkit would be a big piece of that.

Is there anything else you would like to add?

I think the thing you have to remember in this remote work situation is that it's actually a very intimate thing. There's a level of intimacy that is new for a lot of people when it comes to work because they're not used to having their homes be on display. It is doing something really interesting to how people connect and relate to each other that I think is very positive. Often people think that they have their *work self* or *work persona*, but I've always wanted to encourage people to be their whole true selves as much as they can, or as much as they feel they should. I know some of us don't necessarily want to bring all of it home, but this pushes us more to be our true selves at work.

CHAPTER

11

Operations and Administration

When you find yourself suddenly going remote, getting the team set up to continue on projects and to work with clients will be one of the first things that comes to mind. But you must also remember to prioritize the non-client-facing aspects of your operation.

Without the work of the back office, the work of the front office is never fully realized. The company insurance still needs to continue, monthly service fees must be paid, compliance and record keeping must be maintained, and people need to get paychecks. The back office needs to be set up for success so that they can continue the support that keeps the front of the house running. This chapter offers some best practices along with technology that will help keep your business thriving at both ends.

Company Accounts

Not every small business has a dedicated in-house accounting team. Many companies have part-time bookkeepers who will work with outside accountants to keep the company financials in order and maintain records properly for taxes. In some cases, the business owner manages the books and works with various services to stay compliant, while other businesses have a team that handles the bulk of the accounting work for the company. In any of these scenarios, it's best to set your company up in such a way that all aspects of the back-office operations can continue remotely if the need arises.

Automating as much as you can adds to your ability to function remotely as well as helps your teams have more time to focus on more intense work by relieving them of repetitive tasks. Going to touchless payment processing and touchless invoice processing will also help facilitate remote work as well as speed up payment turnaround time. All of these paperless processes combined help promote operational efficiency and better cash flow.

Automate

Many of a company's monthly bills are for fixed amounts or are fairly consistent amounts each month, such as Internet service or rent. Monthly utility bills will fluctuate some but will be relatively consistent on an annual basis. These types of recurring bills should be set up for autopay directly from a company credit card or bank account. Paying bills with credit cards helps free up cash at hand and can earn the company points and discounts from the credit card companies' rewards programs. For example, an American Express Business Gold card will give you points or statement credit on FedEx business shipping services.

In many cases, you can set alerts to let you know if any recurring payments are higher or lower than usual. These alerts can inform you if incentive pricing for new services have expired or your rates change, so you can avoid getting overcharged for services if you are not checking the bills each month. Setting up autopay on monthly bills will take away mundane tasks, but it still gives you oversight that is automated to make sure you are not getting shortchanged.

Make a goal of eliminating paper in your operations and administrative tasks as much as possible. Moving toward using digital payment systems will allow you to take care of administrative work electronically from anywhere without being tethered to any specific physical space.

Paying Bills and Getting Paid

Paying vendors and subcontractors is different from paying recurring invoices. Vendor bills tend to fluctuate much more than typical monthly payments. Cutting checks and sending them out is more difficult when out of the office, but thankfully there are some excellent online services to help companies send out payments and make sure company invoices get paid.

Melio

Melio is a great app for simplifying business-to-business payment processing. It is an easy way to pay vendors and contractors as well as make sure that your company's invoices get paid more quickly. It helps small businesses to automate their payments by making direct payments electronically from bank accounts or credit cards. It even can cut physical checks and send them out for you if a particular vendor requires a hard-copy check.

The Melio platform can set up a payment link for your company that allows you to get paid more quickly and helps you increase cash flow. Instead of waiting for companies to send you a check, you can include a payment link in your invoices that accepts credit cards or ACH bank transfers. This cuts out a step by having funds directly deposited to your accounts. The process also makes your customers happy since it makes the task of paying you much easier for them and reduces their own administrative processes.

Melio syncs and works well with Intuit QuickBooks, which is used by many small businesses. There are no registration fees or monthly service charges, plus there is live support. However, they do charge a percentage for the credit card transactions.

Corpay One

This system automates payments and helps you track expenses to get a better handle on your cash flow. This system will even take care of reimbursements for employee expenses so that your team can get reimbursed more quickly. You can also set up rules for expense approvals to help speed up that process or add in a double approval system if your company requires such approvals under certain circumstances. Their digitized expense reporting makes managing team expenses a breeze.

Corpay One also offers some bookkeeping features. It can help to reconcile accounts to make sure that credit card and bank accounts are balanced against company expenditures. It also helps automate accounting functions by allowing you to create triggers to improve workflow. For example, you may wish to disable autopay if the current bill from a vendor is over a specific amount or if you want to exclude some payable items from being sent to the main accounting system. The system is versatile enough to allow you to create *if this, then that* customized workflows. It also integrates well with most of the major online accounting systems used by small businesses.

Bill.com

This system is a smart accounts payable and accounts receivable automation service for businesses to pay bills, get paid, and manage payments from a centralized cloud dashboard. You can easily manage your company profile and finances from your laptop or from the mobile app. This platform integrates with your online accounting system or it can be used as a stand-alone service.

Bill.com allows you to use your phone to take photos of a bill that is added as a digital document to your payment files. You can also use the system to generate invoices, email them out, and get paid via ACH or credit card. The dashboard is simple and easy to read, so you can quickly see any slow-paying clients at a glance then send them reminders. Bill. com has many online resources available so you can manage how to best leverage their services to better automate your accounts payables and receivables.

BEST PRACTICES

Make a goal to eliminate the use of paper in your office administration and processes. This will cause you to move these tasks online and help you work better remotely.

Online Bookkeeping and Accounting

Keeping your books in the cloud provides you with the convenience of being able to securely access and manage your accounting system from anywhere, which can keep your business going while remote.

All of the accounting platforms discussed here are cloud based and use the double-entry method of accounting, in which every credit into one account requires a debit from a separate account. With this, you know where money is coming in from and where it is going. Double-entry accounting is considered the preferred method of bookkeeping by accounting professionals since it helps to better keep the company's financials accurate.

BEST PRACTICES

Moving your accounting and bookkeeping to online platforms increases your ability to keep your business operational and thriving from anywhere.

QuickBooks Online

QuickBooks Online (QBO) is a major player when it comes to accounting services. Intuit QuickBooks has been a leader in software used by accounting professionals in small and medium-sized businesses since 1998. Account balances appear on the home page of your cloud dashboard to give you an idea of cash flow at a glance. You can easily create invoice templates to help automate your processes. These invoices can be generated and emailed with links to facilitate faster pay via ACH or credit cards.

QBO offers robust reporting capabilities, allowing you to create highly detailed custom reports, or you can choose from built-in report options. Its mobile app has great functionality so you can also keep an eye on finances while you are on the road. The system integrates with over 450 other business apps. These integrations help you cut down on inputting data by having other systems such as sales or inventory sync with QBO. The QBO system easily scales to expand with you as your company grows. At any time, you can add in other services such as payroll or time tracking to keep more back-office processes centralized and automated.

Wave

Wave is a great tool for small businesses and freelancers to keep track of their accounting, and it's free. Users can work on accounting, generate and send invoices, and keep track of receipts and expenses, all at zero cost. The system does charge you for financial transactions. So, when you use its secure portal to receive payments from your clients, you are charged a transaction fee.

The setup wizard makes getting started very quick and easy. The dashboard is clean and intuitive, so it's easy to use even if you've had no prior in-depth bookkeeping experience. For an additional fee, you can add in payroll services. Wave also integrates with Zapier, which allows you to integrate it with hundreds of other business apps to help streamline your remote accounting processes. They also have a program called Wave Advisors that gives a company access to extra professionals they might need to consult with to best leverage their accounting processes. For added monthly fees, you can speak with bookkeeping, tax, or accounting professionals to advise and coach you on best practices for your business accounting.

Zoho Books

This is an affordable and feature-rich online accounting system that can really help speed up your accounting tasks. Its dashboard is easy to use, and you can quickly set up recurring invoices and automate reporting. It is a well-rounded accounting system that offers great reporting and features that would make most well-seasoned bookkeeping and accounting professionals happy.

Zoho also has a full suite of other popular services, such as Zoho CRM and Zoho Inventory, that integrates with Zoho Books to help you streamline your processes. Zoho Books integrates directly with most popular business apps such as Microsoft One Drive, Google Workspace (formerly G Suite), and Slack.

FreshBooks

This accounting system is a good fit for the self-employed and small businesses that are growing. It can handle your needs from invoicing to managing expenses as well as time tracking for hourly services and project management.

FreshBooks will track the invoices that you send and let you know when a client has received them and opens them. It can also send out alerts and reminders for slow-paying customers who are late to help you better manage your cash flow. It integrates with a large library of third-party business apps such as Zoom, Mailchimp, and HubSpot. It offers excellent support via interactive online documents in its help center, email support, and best of all, live operators.

Human Resources

Managing concerns of human resources must also continue as the company transitions to remote work. In fact, HR concerns play a pivotal role in keeping the company on track for success and maintaining its culture.

HR responsibilities include conflict resolution, compliance, and record keeping as well as hiring, onboarding new team members, and saying goodbye to others. When the company is remote, these tasks rely heavily on excellent communication, effective processes, and the right technology.

No Dedicated HR

Many times, small businesses don't have any dedicated HR staff. This means that there are employees in the company who wear several hats, and taking on responsibilities of human resource personnel is one part of their role. That can be daunting under normal circumstances. When the company is forced into a remote work situation, wearing all those hats while managing the switchover can be highly stressful.

If you find yourself in that situation, then delegate those HR responsibilities to share the workload and relieve the burden. Avoiding burnout is important and something you should proactively combat. Along with delegation, make use of technology to help automate processes and enable HR self-service. HR self-service allows company employees direct access to their most common HR needs via an online portal and dashboard. Among other features, it allows them to view and make changes to benefits and adjust their payroll configurations. This removes that workload from those who would normally manage those HR responsibilities and it saves hours of their time.

Record Keeping, Compliance, and Benefits

Tracking employee performance is part of your company record keeping. Although you may see performance, measure it via employee engagement each day, and offer constant feedback; performance observations should be recorded for the purpose of evaluations. Any disciplinary actions also need to be reported and filed.

Other human resource tasks that need to be considered are medical benefits, taxes, labor laws, and benefits administration. There are excellent cloud-based services that can help alleviate the burden of HR management and keep that work flowing at all times. Some key software features to look for are employee self-service portals and benefits management.

BambooHR

This online HR service is great for small businesses and is very easy to use. One of its more powerful features is the employee self-service portal, which helps to eliminate some administrative steps. It allows employees to update their personal information, put in time sheets, request time off, or access HR documents. It also allows team members to leave anonymous peer reviews. This gives management the ability to gain insights on employees from the perspective of the co-workers and give those employees helpful feedback.

BambooHR has a strong applicant tracking system that provides visibility on every stage of the hiring process, from the initial application to making someone a job offer.

The mobile app is a strong asset for getting company info at a glance, such as their Who's Out Calendar, which lets you quickly see who has requested time off that day and won't be working. Managers can easily input performance notes and collaborate with each other on job applicants and new hires.

The mobile app is robust, and the whole platform integrates with many third-party apps that you can find in their HR marketplace. Notably, it connects with benefits companies such as PlanSource and Maxwell Health to help employees with benefits enrollment and management.

Zenefits

Zenefits has strong features for employee benefits administration along with HR task management, and it keeps adding new features to do much more. The hiring and onboarding app integrates with the main Zenefits system to make processing new hires streamlined and efficient. It can easily run background checks, send offer letters to new hires, and allow them to get the onboarding process started through an employee portal. This is all done digitally to help you eliminate paper as much as possible.

You can add on other services such as payroll and professional advisor services to help you with more detailed HR needs. You can also add in your own benefits broker for a per-employee fee if you happen to use a broker that is not within their list of benefit partners. If you don't have a broker for health plans, then you can shop their Zenefits Health Insurance Marketplace where you can find and compare coverage in your local area for nearly any size company.

Namely

Namely is an online HR platform that is highly customizable as well as feature rich. The portal allows managers to set up a news feed that will inform team members of upcoming events, policy changes, or other info when employees log in. The feed can also include team members' birthdays, work anniversaries, and other milestones to help maintain engagement and socializing. Employee profiles can be set up to include skill sets, professional goals, and brief bios. The system acts as a central repository for HR docs such as company policies, benefits, forms, and other info that employees should have access to at any given time.

Most HR software allows for roughly three levels of access—for managers, administrators, and employees—but Namely is fully customizable. A company can enter its own hierarchy within the software to allow different levels of access and capabilities for any level of leadership, from shift leaders to CEO, and for any department. The system features strong employee onboarding capabilities, and its configurability allows for a high level of customization to suit the needs of any industry.

Like other HR software, Namely offers a mobile app, payroll and benefits management features, employee engagement, and HR compliance. It has numerous integration partners, including outsourced HR advisory groups, savings and retirement companies, and employee training solutions.

Gusto

This company started off as Zenpayroll and expanded its service offerings to help small businesses navigate and manage more HR needs, such as benefits, time tracking, hiring, and onboarding. It is also a company that is driven by its values. Some of its guiding principles include "Do what's right; what's right isn't the same as what's easy" and "Go the extra mile; go beyond delivering what works; discover what delights."

The system offers hourly time tracking for employee payroll, hiring and onboarding features, and management of health and other benefits such as 401(k) accounts. Its concierge service will connect companies with certified HR professionals to help organizations stay compliant and have quick access to resources that help them get the answers they need from HR experts.

An impressive feature Gusto offers is Gusto Wallet. This service helps employees get the most out of their paycheck by helping them plan for the future and manage their expenses. They can set savings goals, easily change their contributions, and route portions of their paychecks to different bank accounts such as savings, checking, or vacation accounts. Should an employee have unexpected expenses arise, they can even access some emergency funds between paychecks with a paycheck loan that carries no interest and no fees.

Paycor

Paycor is an excellent all-around tool for the back-office administrators. It offers features such as payroll, benefits management, talent management, workforce management, and employee experience management. This means everything from the hiring experience to scheduling to employee engagement can be managed via the online portal. This system is so robust, it is considered a human capital management platform as opposed to a human resources platform. This software puts an emphasis on features for employee retention as well as ongoing professional development. The system is scalable and can work well for companies of all sizes.

With Paycor, everything is centralized on the dashboard. Employee self-service portals are easily accessed from mobile devices. Paycor helps to streamline HR processes by automating workflows and repetitive tasks while it helps admins to stay on top of things with reminders and notifications. The dashboard is very user-friendly, which helps to facilitate employee use.

BEST PRACTICES

Small businesses can benefit greatly from cloud-based HR platforms to streamline their human resources management.

Losing People during the Switch

Remote work is not for everyone, and it's better to find that out early on if there are existing team members who will not thrive in a WFH scenario. Some people simply work better and are more comfortable with the in-person camaraderie of the traditional office environment. A number of the professionals that I interviewed for this book did lose people shortly after their companies went remote. Some people left for personal reasons, while others left for new opportunities. However, some of these voluntary exits were quite revealing.

Some of the leaders mentioned to me that they lost employees when they first went remote in part because of the new methods of gauging performance that needed to be used when the team became distributed. These individuals could not handle the self-scrutiny and independent work that comes with being remote. It turned out that going remote highlighted their general lack of performance.

Being in an in-office and in-person collaborative environment allowed many underperformers to fly under the radar because as their teams completed projects, they would reap the reward of the work largely done by others while contributing very little of their own. While employees are working remotely, ideas and contributions are easily traced through their digital footprints in the form of emails, messages, or whiteboards in various apps, so the people who are the most active in a project really stand out more individually.

Remote work can flip things upside down, and you may see some people shine who you previously saw as less engaged and underperformers. Some leaders shared with me that people who are normally quiet and did not offer much input during team meetings around the conference table were suddenly speaking up and were active contributors during virtual team meetings. It turns out that some of the more extroverted individuals on the team had big personalities that would drown out these other team members, which resulted in their lack of participation. But working from their own spaces, those withdrawn individuals were more comfortable and able to be more fully engaged with their team.

Remote work can be a double-edged sword in that way for employees. It will be great for some and a poor fit for others. The experience may help weed out employees who may not have been a good longer-term fit for your company at the same time that it reveals some stars you hadn't realized you had.

Remote Hiring Process

Hiring people remotely is quite different from the traditional way of recruitment. Remote hiring requires a different style of interviewing and different questions for candidates as well as a different way of onboarding. You may realize that you discover some methods of hiring remotely that will work even better overall than what you normally use for your in-person processes.

One of the business leaders I interviewed discussed some of the differences in her remote hiring process compared to her in-person hiring process. Here's what Leslie Murphy from Raybourn Group International had to say:

We hired and onboarded four new people remotely during the full lockdown as well as onboarding someone we hired right before the middle of March. Here are some things we learned in the process:

- We utilized Zoom for all interviews and advertised we were doing that to help people feel comfortable about the process.
- We will always utilize Zoom or Teams in the future, at least for first interviews. The benefit was that we were able to engage with more of our remote and non-Indianapolis-based staff, which was very beneficial to the process.
- Since so much of our work is done virtually, we utilized the interviews to gauge a candidate's familiarity with technology and how they presented themselves. This would have been more challenging in person.
- We had to adapt our usual in-person skills testing to a virtual environment. We will keep that moving forward as we can skills-test candidates as another way to assess if we should interview them.
- For onboarding, we focused on a high-level of virtual interaction to engage, educate, and evaluate how each new staff member was progressing. We kept our usual 8- to 12-week onboarding schedule but instead of the first week including how to use the office equipment, we were able to focus more on understanding our company, their role, and learning more about the rest of the staff.

As you see from Leslie's experience, when hiring remotely, there will be some adaptation to how you normally would hire. But these changes are good. They help broaden your company's abilities in hiring and increase efficiency in your operations.

It Begins before the Hire

Getting the right team members is extremely important for maintaining your vision for the company and for helping your leaders maintain the company culture. This need is so great that many companies incentivize their own employees to refer people for new members of the team. Among the benefits of an employee referral hiring program is that this word-of-mouth recruiting will come through someone with first-hand experience with the company who can give an honest account of the work environment and what the job entails. This practice increases the number of quality candidates and increases employee engagement. It also helps the company to better attract talented people who are a good cultural fit.

You should make a point to showcase your culture in your job postings. Make that a strong part of your recruiting strategy. You will filter out applicants who are not the right fit by being clear and upfront about what is expected in the attitude and outlook of team members. Use language that promotes the higher purpose of the job. If it's customer support, give a job description that showcases your vision for the company as well as how a person in that position betters the company. The following excerpt from a public job posting for a customer support representative from San Francisco–based Aurora Solar is a great example of incorporating the company culture into your recruitment strategies:

> As a Customer Support Representative at Aurora Solar, you will ensure all our customers are set up for success by answering product questions, providing best practices, and assisting the customer with any technical hurdles that may arise. You will also influence our product design and direction by relaying customer requests back to Aurora's Engineering and Management teams. If you enjoy working with people, and geek out about the technical and business aspects of the solar industry, we would love to hear from you! We are a diverse team based in San Francisco, passionate about advancing the growth of solar energy.

A more traditional posting may have said that the applicant will be expected to work from 9 to 5 during weekdays to answer calls, emails, and chats from customers regarding solar products. That may get someone to fill the seat, but it would not be the best strategy to get someone who fits the company culture.

Let's take a look at some of the nuances of this posting in how it drives company culture and purpose-driven work. At its core, this posting is about being in a position to help outside people as well as to improve the company. It's a job that is positioned to help drive the success of others while also promoting the higher ideals behind renewable energy. In another part of the posting, the ad takes the time to include part of the mission and vision of the company.

> About the role: Based in San Francisco, Aurora Solar is a fast-growing Series B company focused on building a clean energy future for all. Aurora makes the software that is enabling society to transition to a world powered by solar power. We want every solar installation in the world to pass through our software, which has already been used to design millions of solar projects.

> Named "Top 50 Tech Companies to Watch in 2020" by BuiltInSF.com and awarded #1 Solar Software platform by Solar Power World, Aurora is disrupting the energy industry and changing the course of history. We are experiencing hockey stick growth and are looking for A-players to come join and accelerate the fun. We are looking for a motivated and detail-oriented person interested in joining an organization that is making a meaningful difference in the world.

Including this additional information is a good strategy to get the best candidates. You don't want just anyone for a job, you want the candidate who will fit best in your team. In your postings for a job, be clear and authentic about who the best fit is for a role as a team member in your culture. Share your vision and your mission. By including these aspects of your company, you save aggravation. Hiring the wrong people for your company culture costs you far too much money, not to mention the time wasted training them only to have them leave because they were the wrong fit, putting you in the position of having to again begin the process of searching for the best candidate.

All Digital Engagement

Keep in mind that when you hire remotely while your company is working from home, the new hires will also probably be out of the office when they start work. You may not meet them for weeks, for months, or possibly ever. This means that for a good amount of time, nearly 100 percent of your engagement with these new hires will be digital. This can be an advantage to your hiring options. If this is someone that you decide will be a fully remote employee and will not have need to come into the office, then you are not limited by geography in your search for a good addition to the team and that can be a great boon to your company.

Take advantage of a much larger pool of talent and try out some of the national job posting and recruitment sites such as Glassdoor, ZipRecruiter, and LinkedIn. Or you may want to try out some of the sites that are dedicated to remote workers such as Remote.co, FlexJobs, and We Work Remotely. These sites are great sources for candidates for full-time, part-time, or freelance work that will take place remotely.

Questions for the Remote Worker

You will find that you need a slightly different style of questions for interviewing the potential remote worker. Just as management styles must adapt from the in-person office experience, it's necessary to adapt for recruiting a remote worker. The questions you ask will get more specific for various industries and your business needs, but I have compiled a short list of topics that tackle some of the unique challenges of working from home. Use these to address those challenges and find out how the potential employee would ease into their new role and work without having someone a few steps away to help them as they would in a traditional office setting.

Remote Work Experience

It is not a deal breaker if they have no remote work experience. Before the pandemic, the overwhelming majority of the global workforce had no remote work experience. But you should ask the question to better gauge the needs for their adaption to the WFH environment. This will help you to identify where they may have some struggles and allow you to create a more defined road map for their success if hired.

This is also an opportunity to ask why they want to work remotely. It may be that they want to have more time with their family, they may be the primary caregiver, or they may live in an area with fewer good job opportunities and don't want a long commute. This helps you better understand motivation as well as their personal home situation. With flexible jobs, it's important for employers to understand these nuances to promote better empathy and engagement with their team.

Measuring Success

Many companies with a traditional workforce measure productivity and success in an employee's work very differently from how they would with a distributed team. Make sure your candidate is aware of how you will be gauging the success of their work. It may be something that they are not accustomed to, and not all potential candidates may be comfortable with your methodology.

As we have discussed, remote work is not for everyone. Some people may think that being able to work from home sounds like a dream job, but when they find out how that work will be measured and some of the things that will be expected of them, they will realize that it may not be for them. Remote work relies on a good amount of self-discipline to create and maintain a high quality of work. There is far less room for hand-holding and personal supervision to ensure good work is produced.

Ability to Maintain Focus on Work

It may seem like a broad question, but you should ask what the candidate does to keep their focus on work. While working from home, they may have pets, kids, roommates, or other distractions that would not be present in the traditional office. They will need a plan to ensure that they can maintain their work regardless of distractions. Their responses may allow for you to ask more specific follow-up questions about what they may do to keep focus in certain situations that may impact remote workers. If their house is overrun by afterschool activities, they may say that they have a nearby coffee shop or other space that they can go to and remain focused on work. This can be a great segue to the next question.

Their Workspace

Asking a potential employee about their workspace helps you understand the environment where they may be doing the bulk of their work for your company. You may need to make some suggestions for them based on content from previous chapters to alter their workspace to combat fatigue, eye strain, or other remote work hazards.

This also opens the door to find out what they may need in order to be a more productive and fully functioning member of your team. You may want to invest in better broadband and add a reimbursement for the higher monthly fees. You may want them to have a better computer to work from or even a better chair that is more ergonomic to prevent neck and back issues.

The Biggest Concern with WFH

Ask them what their biggest concern is about doing their new job while working from home. Again, this will be very revealing to their unique situation and will allow you to take steps to mitigate or even eliminate their concerns about the job.

This also will tell you how much they have thought through their new work-from-home position and if they have assessed any potential pitfalls they may encounter. You want to know if they have identified any hiccups that may come up and if they have thought of a way to deal with these potential problems. This speaks to their ability to plan ahead and problem solve as well self-manage their work day. These are all critical skills in the WFH environment.

Company Values

Find the wording that works best for you, but you need to ask if their values align with your company values. That sounds direct, and it should be—for an important reason. It is vital to the success of your company and your team for you to hire the right people. The right people are those with the needed skills and whose values align with your company values.

The best candidate has to be a good cultural fit. That culture is founded in the shared values of its people, and those values drive the everyday work of the company. If a candidate's personal values do not align with the values of your company and its culture, then that is a definite deal breaker and you need to move on to the next candidate, even if for a temporary or seasonal position.

Remote Tech

Ask what remote work technology they may have used in the past. You will train them on how to use the technology that your team uses now and how to use it in the manner that you feel appropriate, but you should get a feel for how comfortable they will be when they are fully immersed in remote technology for daily work.

You also will want to have your line of questions set to understand their technology skill levels. This will give you an idea of how adaptable they are to new software platforms and how to troubleshoot them. Unless the job calls for it, you will not need someone who can code or build web pages. However, you do need people who can use and understand cloud storage for files and documents as well as online communication tools.

When the team is distributed, they will not have easy access to IT personnel. Should a new employee have an issue with some of their tech when a meeting or deadline is looming, how they will act and react is important. If they tend to be stymied easily by technical issues or are accustomed to having someone more tach savvy than them resolve their issues for them, they may not be a great candidate for remote work.

If a candidate is not familiar with the tools that you have in use, they should at least hold high value on learning new skills, being open to feedback, and being resourceful. They should also be willing to tackle problems on their own. This is not to say that they will be left to learn the programs on their own, but these traits will be useful for the remote work they will be doing and for their ability to adapt and resolve issues independently.

Attributes and Skills

Regardless of how long an employee may work from home, remote work requires a different set of skills than those used primarily when working in person. Certain work habits and work styles that may only have a marginal impact on productivity in the office can have a huge impact on remote work and remote teamwork.

A good candidate as a remote team member will be able to remain focused and productive with minimal supervision. Self-discipline and maintaining motivation while working alone are essential to their success. During the hiring process, team leaders should look to assess the candidates for the skill sets such as the ones discussed in the following sections to make sure they are hiring people who are up to the job of remote work.

Written and Verbal Communication Skills

You will get a glimpse of a potential employee's written skills when they respond to the job posting and their verbal communication skills during interviews. As we have discussed in previous chapters, in the remote work environment, it may be more difficult to get a point across effectively than it would be if you were making the same point in an in-person office environment. As has been reported, teams that were new to working remotely might take several meetings to get on the same page, whereas when they were in person, a single meeting would have sufficed.

Keep in mind that a large amount of the new employee's interaction with the rest of the team will be in the form of chats and asynchronous communications such as written notes on online project boards, task updates, and emails. An exceptional remote worker will be more precise with their use of language in their overall interactions with the team. That means being able to be direct, concise, and specific verbally and especially in written communications.

Being a Self-Starter and Being Independent

A strong remote team player needs to start the workday on their own and be able to self-motivate to keep up their productivity throughout the entirety of the day. There will not be a manager popping in or giving hourly check-ins. A remote worker who can thrive in a WFH situation will be able to maintain productivity on their own.

If the person has ever freelanced or worked independently in the past, they may be an excellent candidate. People who have done this type of work tend to be able to adapt well to change, to manage unforeseen hiccups, and to be resilient. Knowing when to ask for help and being able to ask for help are also important traits for good remote teamwork. Some people may be unwilling to ask for help due to fear of seeming incompetent, or it may be an ego that demands that they figure things out for themselves. This can have a negative impact on the rest of the team and slow them down. Ask them how they would know when to keep trying to figure out a problem on their own and when they should stop to seek additional assistance from a team member.

Organization and Time Management

A great remote worker will be able to establish a routine, get the info they need, and create a plan for attacking the work assigned to them for the day and the week. In order to be able to do that, they must keep themselves well organized. People with clear organizational skills are able to maintain focus and stay motivated to get their work done in the necessary time frames. Everyone has their own styles to keep themselves organized. Be sure to inquire about the candidate's organizational methodology.

Ask them about their personal approach to time management. Being able to budget their time on projects through the workday will be a key component of their effectiveness as a remote worker. Part of this is the ability to prioritize. They need to be able to recognize parts of a project that are important or time sensitive and know to focus on those items first. Knowing which tasks take precedence and when to stop working on less urgent tasks are foundations to remote a worker's time management and organization.

Reliability

You are asking someone new to be able to self-manage and perhaps make their own hours. It's an understatement to say that you need someone who is both reliable and trustworthy. You are entrusting your business, your clients, and the hard work of your team with them.

Find out if this will be the only job they are working, what challenges they may have when working from home, and what issues they may foresee during the work week that may impede their ability to do the job. Be straightforward in this line of questioning. There is no need to be overly subtle about this issue.

You can also ask some interview questions that are designed to help you better gauge the candidate's reliability and work ethics. These types of questions help the candidate to pull from their personal experiences and express themselves more than they can with simple yes or no style questions. Here are some examples:

- *Tell me about a time when an important personal issue came up during work hours that you needed to deal with and how you handled it.* We will all have personal issues come up that pull us away from our tasks and work. The answer to this question will offer some insights as to how the candidate will manage themselves in a difficult situation and without direct supervision.
- *Tell me about a time when the workday was over but you still had some unfinished project tasks that needed doing.* The answer to this question will help you gain insights to their work ethic and their commit-

ment to being a reliable team player. Missing deadlines and unfinished tasks will stymie the work of others on the team. A good remote worker will strive to take the needs of the team into consideration when they make decisions.

- *Tell me about a time when you were assigned a complex task but had little supervision and little direction on how to complete it.* In a remote environment, communication can sometimes be haphazard, and getting the intent of a message across to coworkers may not be as clear as we would like. Not asking for help due to feeling embarrassed or fear of seeming less capable will hold back remote teams. It is important for a strong remote worker to be willing to reach out to peers and managers to get further clarification and more direction when needed.

Collaboration and Teamwork

In the team environment, each member brings a piece to the table that fits with the other pieces. Each member's work is dependent on the work of others. Along with the other attributes previously listed, being able to listen well, ask questions, and take direction are critical to being a strong remote team player. Ask about the candidate's teams in other places that they have worked to gauge how well they have collaborated in the past. Ask them about the strengths and weaknesses of their previous teams and their own strengths and weaknesses as a team member. Also ask them about a time when they had an argument with a co-worker and find out how they resolved the issue. This will help you gain some insights as to how they resolve conflicts and will tell you more about their interpersonal skills.

Work-Life Balance

Ask the candidate how they unwind and how they switch off from work for the day. They need to be in a remote work mindset and part of that is being sure to keep that healthy balance between work and home. If they cannot build those boundaries, they may be more prone to suffer from burnout. It happens frequently with people who have strong a work ethic and who tend to be workaholics. Some employers prefer those types of work styles, but they are not optimal for remote work.

Ask about their interests outside of work. Find out about their hobbies and activities. What leads up to that burnout that you are trying to avoid is not just the imbalance itself, it is the resulting stress, unhappiness, and disengagement from the work that then leads to looking for other opportunities. You can of course take steps to mitigate those issues once you hire a new employee, but you are looking for the best candidate to add to your team. Ideally, you want someone with those attributes instilled already, unless of course you are looking to be a great mentor and help that with that aspect. If so, that is a fantastic ideal to strive for, but be sure that you are aware of the level of personal engagement and time that you will be investing. Keep in mind that you are holding their career as well the rest of the team in your hands. Be sure to bring on the best fit and not try to force-fit anyone.

BEST PRACTICES

You must adjust each stage of your hiring processes to be specific to the company's remote needs in order to hire the right remote team member.

Remote Onboarding

Some people think of training and onboarding as the same thing, but onboarding is different from job training and company policy training. Onboarding is introducing new hires to their immediate team members, your executive team, and other departments as well as helping them adjust to the company culture, values, and expectations for their role in the organization. On top of this, you are also setting them up with HR, payroll, and online profiles for communications and collaboration such as email and Slack.

In the traditional office, a tour of the facilities helps with many of these things, but that may not be possible with the remote workforce. During the remote onboarding process, the new hire will be seeing lots of new faces, and they will get a better understanding of the processes that make the company work, how its workers are supported by management, and the values that are the basis of your culture. All of these aspects of the company need to be relayed remotely.

Make a Solid Plan

You may want to look at your onboarding plan that you have had with traditional in-person work and tweak that to meet the current remote needs and then add a few new steps. If you normally have a new hire shadow someone for a week, that can still be done with some technology, but it may not be optimal in all cases.

The onboarding process will be among the first experiences this individual has with your company and the team, so it will make a big impression on a new hire. They will be assessing your company just as you are assessing them. Make sure your new remote onboarding process has the same elements and high standards of quality as the rest of the work your company does.

Solicit Feedback

Get input from the new employee regarding how things are going during onboarding and training. This is new to them, but bringing on a remote person may be new to you as well. There will be bumps in this road, and being new to it, you may not see them all before you hit them. Besides better engagement, this is another good reason why you should get feedback and act on it if needed.

You also want new employees to feel valued. Soliciting feedback helps them to feel like a valued new team member. Transparency is important to a strong remote work culture, so don't be afraid to let them know if they are your first remote hire.

Facilitate Social Connections

Another part of the onboarding process is for you to help the new employee get to know and start relationships with their new team. It's important for strong teams to build emotional and personal connections to each other. You can't just ask to bring them to the front of the office and have them introduce themselves to everyone. You could do that on a group video meeting, but that's impersonal and can be intimidating to the new hire.

Help them recognize who is on the team. Share an organizational chart and go through employee profiles on the various communication channels as initial steps. Take the time to do one-on-one introductions and then group introductions.

Getting the Right Team Members

As Trevor Yager, CEO of TrendyMinds, says in the following interview, the right team members are essential to the success of your company's remote work. Once you have that team put together, you must keep supporting them in the work that is done on the client-facing side and the non-client-facing side. Good technology backed by good remote processes and automation will help keep all aspects of the company healthy and keep up the job satisfaction for your team. Thinking in those terms is part of a company culture that puts its people first.

TREVOR YAGER, PRESIDENT AND CEO TRENDYMINDS

Company profile

- Location: Indianapolis, Indiana

- Employees: 75

- Primary Line of Business: Film, marketing, Web

- Primary Audience: Healthcare/pharma, tech, and other enterprise clients

About us

TrendyMinds is an audience-focused marketing agency that helps organizations share their message with the people who matter most. As consultative partners, we immerse ourselves in our clients' brands to provide creative solutions across film, marketing, and Web.

What tech services/software did you use to go remote?

Slack, G Suite Business, TeamGannt, Harvest, Basecamp. (We already used these, though, so they weren't implemented just to go remote.)

What was the most difficult part of going fully remote for you and your team?

Most of our services were able to continue with very little disruption. However, our film division saw many shoots postponed due to social distancing and safety regulations.

What was the easiest part of going remote for you and your team?

We have had remote team members since 2012, so we already knew how to work together remotely.

How did everyone working remotely affect your team working together? Specifically, were there differences in the generational groups when they were put into the situation of suddenly working remotely?

We didn't notice any differences between the generations; however, we did notice differences based on personality type. For example, some people really crave in-person, face-to-face interactions while others flourished by working from home. We also noticed additional stress on those with children as they had to become adept at schooling in addition to parenting and their career-related tasks.

Did you notice much difference in how your team worked together when remote?

Yes and no. We noticed an uptick in efficiency as people were able to focus on their tasks, and meetings were a lot more focused. We did have difficulty with brainstorming meetings or those that we would typically hold in-person. However, the team learned to adapt to video calls in order to accomplish these tasks.

Did you create any avenues or methods for your people to stay social with each other?

Yes. We created new Slack channels surrounding various interests such as gaming, sports, movies, and *The Bachelor*. We held many happy hours to celebrate successes and welcome new people. We also had a few official company lunches where everyone was able to order lunch via Uber Eats, and special swag kits were sent to every team member's home. In addition, we encouraged teams to socialize outside of work via online gaming or socially distant activities like walking or bike riding.

I also wrote a weekly note to the company updating them on our company financials, client projects, social activities, helpful tips, and just encouraging them to take care of themselves and those they love.

How did you work remotely with clients?

We had a lot more video meetings than usual. By not having to drive to meetings, we found more efficiencies with time management. We still used email, Slack, and Basecamp to communicate project status as well.

What would you do differently?

I'm sure there are things we could all do to improve, but this was definitely a first. I think I'd encourage our team to try to have a bit more fun. There was a lot of doom and gloom, and rightfully so. But I think we could have done more to encourage people to seek positivity in life.

What advice would you give?

Surround yourself with great team members. We would have failed if it weren't for our people. No amount of planning or tech tools would save a company if the team was subpar. I feel very fortunate to work alongside an incredibly talented group of individuals.

The Complete Package

When I was a boy, I used to dream of people working remotely. That may sound a bit odd, or even unbelievable, but hear me out. I'm a Gen Xer. I was one of those latchkey kids I mentioned in Chapter 9, "Generational Struggles." Both my parents worked, and my dad worked two jobs. One of his jobs was at JFK Airport and the other at LaGuardia. Both jobs were as a facilities mechanic for TWA and American Airlines. With Mom and Dad both working, that meant that I spent a great deal of time home alone with my brother, who was two years older and my only supervision.

Being home without parental supervision may sound like a dream to some kids, but after a while it gets lonely. It also got isolating because we were told to stay home and not open the door for anyone. This meant no social interaction with kids our own age other than at school. On birthdays I would make a wish that we would have some life-changing event that would allow my folks to be home. Maybe we would win the lottery, or we would receive a large inheritance from a long-lost relative so that my parents could quit their jobs.

Then I started to think of ways my dad could do his job from home. In my mind, I invented all sorts of robots that could be controlled remotely over long distances. I imagined my dad hooked up to a contraption in the den where he would be able to fix things robotically at the airport through what would today be called telepresence.

At the time, I was determined that when I grew up, I would go into robotics so I could create these machines and help kids spend more time with their parents. Life didn't pan out that way. Instead, I helped to develop companies that used SaaS technology, telecommunications, cloud-based dashboards, and centrally managed systems to remotely help other businesses become more successful. With that success, it is my hope that these business owners are empowered to be able to spend more quality time doing what they enjoy and with the people they love. I hope this book and the information within empowers readers in the same way. Let's do a quick review on how best to do exactly that.

Work Remotely with Added Confidence

After reading this far, if something happens that makes you need to shut down your office and take the company remote, you should have a strong idea of the next steps to take. It may not be a crisis that causes you to have your employees work from home. At some point you may make the strategic choice to transform your company into a virtual organization, or you may create a hybrid of a staffing structure comprising of remote workers and those in the traditional office. In any of those scenarios, you are now well armed to be able to not only keep the company operational but also to thrive while you are remote.

Even though companies such as IBM and Yahoo! had once reversed their work-from-home policies, the migration to remote work due to the COVID-19 pandemic proves that companies can indeed thrive when many members of their team are not in the office together. If anything, the COVID-19 shutdowns have given us a massive amount of good information on working remotely, and it has highlighted the pitfalls to look out for along the way.

Clearly our remote work technology today is more advanced than what was available to IBM and Yahoo! in 2017 when they called for the mass exodus back to the office. And even though companies in 2020 still reported issues collaborating, technology and company cultures quickly stepped in and stepped up to alleviate those frustrations. They were able to use new findings to make additional adjustments to circumvent the disconnects in getting the message across the digital divide and increase collaboration efficacy.

We now have a deeper understanding that the isolation that was reported by so many can indeed be prevented through better engagement and increased transparency. Steps can be taken to mitigate video meeting fatigue and burnout. It was because the business world depended on resolutions to those issues that crippled productivity that methods to alleviate these problems were found. It reminds me of the quote by Euripides, "Nothing has more strength than dire necessity." In 2020 we did not have the ability to choose to go back to the office as companies could do in 2017. The business world *had* to find a way to continue to thrive while remote, and so the business world evolved to make it happen.

One of the takeaways you should have is that the better prepared you are for an event that forces you to take your company remote, the smoother your transition will be. You need to have a continuity plan for your company, and the bones of that plan need to be installed. With the right office systems and software already in place, you will be able to maintain productivity the same day your office shuts its doors to send everyone home to work. Getting your company and your team started on remote-friendly technology should begin right now.

Make Your Space Your Own

Your home office should be rearranged strategically to meet your needs and keep you focused. Be sure to find ways to segregate your work area from the rest of the house if at all possible. Create boundaries for your workspace and your work time so family and friends know when you are doing your best to stay in the right frame of mind for the job.

You now know what hours of bad lighting and sitting in a poorly designed chair can do to both your body and your mental state. Don't take the need for these comforts too lightly. If you are the type of person who tries to power through adversity and suck it up, then you need to give yourself permission to invest in the furniture and home equipment that you will need to keep up a good level of productivity.

If you are a leader in the company, then you need to advocate for your team to ensure that they get what they need for their home workspace as well. These aren't niceties or luxuries. They are needed tools to fight against burnout and both mental and physical fatigue. They are productivity enhancers, widgets to maintain motivation or whatever else you want to call them to convince yourself to make these purchases as investments into your employees and the business. Have the wisdom to know that they are indeed necessary for your company to continue to do good work when working from home. Make your well-being and the well-being of the team top priorities.

Getting into a daily routine will prime your mental state for your workday and your workweek. Make getting dressed for work a part of that routine. It sets your frame of mind into the right mode and makes it easier to shut down when you change after work, giving you a better balance in your home life and your professional one. Setting SMART goals (which you'll remember from Chapter 2, "The Remote Workspace: Set Up Your Mind and Your Space"), for yourself and for the team, will keep you on track and productive. When these goals are combined with strategically timed and enjoyable breaks, you will be able to keep up your momentum. Remember to replenish your personal resources early, before they get depleted, so that you have the mental and physical stamina to last the entire workday.

Team Strategies

Maintain productivity with the rest of the team by being proactive in communication and through extraordinary clarity on your expectations. Be sure the team fully understands what communication tools are appropriate for the various types of conversations that come up. Having detailed remote communication policies available in a public or shared document is a best practice that will help keep everyone on the same page with little room for misinterpretations. This is especially needed when it comes to asynchronous forms of communications. Having clear, well-documented communication policies will help save time and frustration for the team.

Part of your remote communication strategies should be the daily check-in. When the team is distributed, the daily check-in keeps everyone on track and helps add structure and normalcy to the workday. It is helpful not just for the sake of productivity but also to combat social isolation. Staying connected in this manner is a powerful source of vitality for the remote team. It makes being apart from each other easier to deal with, and it refreshes the idea that the work being done when working alone remotely is just part of the larger work being done by the rest of the team.

Part of your remote team communications strategy needs to be making yourself available. Showing yourself online and accessible to the other workers keeps the remote work flowing smoothly. It also keeps the door open to socialize or ask quick questions for clarity and provides some comfort by just letting people know you are there for them. This methodology allows you to be much more than just a face on a screen. It enables you to create and maintain real connections with your people and better interpersonal relationships with the distributed team members.

Technology Is Your Asset—Choose Wisely

Plan for tech that best helps you to keep working with both the team and your clients as well as potential clients. Make strategic plans and communication polices to cover internal communication needs and various external communication scenarios. Create chat channels to cover your anticipated topics and departments as well as channels to have social interactions. This will help keep the team organized and maintain productivity.

Take steps to ensure there is clarity in how tech is used and for which needs. Excessive emails and video meetings kill productivity and drain mental resources. Endless email threads or chats taking place in channels where they don't belong can also be sources of irritation with distributed teamwork. That sounds like a small thing, but many small irritations can build up and increase the chances of diminished engagement when remote.

Taking the time to set up documents in secure cloud storage now will help keep work moving when it comes time to go remote. It will minimize any disruption in your workflow. Because your team will all be logging into centralized systems from various locations, you will want to ensure that you keep your security up to snuff on all ends. Protected Wi-Fi and use of strong passwords and double authentication are important first steps in any cybersecurity initiatives. You will most likely choose to layer on more security methods such as VPN as you judge what you need to keep your company data and client data safe from prying eyes. Make cybersecurity a strong part of your continuity planning.

Scalability and TCO

It will take some serious strategic planning to acquire and implement the various technologies that best suit your needs. If budgets are a top issue, then think long term. It costs you more money to start off with equipment or software that is less expensive but doesn't fully meet your needs. You will end up needing to switch to something else not long after. You will then lose your investments in the old tech, the installation, and the time it took to get everyone set up and trained.

When reviewing any office technology, look for those that can scale up to meet your needs as you grow. Take a look at your growth plan and let that help to guide your choices. Future-proof your purchases as much as possible by looking at the road ahead and letting that set your direction. High performance and reliability should be up at the top of your checklist when choosing providers. Compatibility with your other tech should also be a factor. When multiple software platforms integrate well with each other, you reduce the repetitive tasks of entering the same info into multiple databases and free up time for deeper work on more important tasks. The point of office tech is to make your work more efficient and reduce menial labor. Leverage it that way to get the best remote work experience and reduce stress. Explore the features being offered and get as much automation out of them as you are able. Don't be shy about asking for features that might meet your future needs. Many of my own companies' service features were added due to customer input and requests. SaaS companies are very competitive and constantly look for new features to add. Getting features requests from clients helps ensure that new software abilities get added on more quickly in future updates.

Regardless of the type of software you might be looking to add to your company, look for your total cost of ownership. Now that you know what to look for in ongoing and initial costs, have your project manager get the full info on what it takes to set up and get going. Being able to get going on your own or needing to pay for someone to help onboard your company is a big consideration in the equation of time and costs. Finding out what kind of maintenance needs as well as the level of service and support you can expect is also something to keep in mind during the vetting process.

Make the Most of Your Time

Time management for a newly remote worker can be tough. If it happens during some crisis that forces you to go remote, it will be even more of a struggle. Take all of that into consideration when looking at team productivity as well as your own. Create new methods of measuring productivity that are results based and goals based.

In stressful situations, it's easy to get distracted and even easier to fall into the trap of productive procrastination. You have an idea of some productivity apps to help you reduce distractions and assist you in enabling deeper work. I can tell you from personal experience that they do help. Use the daily check-ins to help your team keep their focus and prioritize their tasks. Look at goals as well as deadlines and leverage the check-ins to adjust workloads and target dates accordingly. As your team adjusts to their new work-from-home situation, these benchmarks and regular status reviews will be immensely helpful in reducing stress, keeping the team organized, and setting them up for success.

Adding some helpful programs to maintain your efficiency will be a big asset for your mental clarity. The use of calendar apps can help with organization and scheduling. Introducing automation from app integrations will be huge for time management and reducing burnout. Allowing these apps to handle repetitive tasks increases team workflow and reduces the drudgery of doing the same thing over and over. That alone can be a big factor in combating burnout.

Keep the Faith

Trust can be a source of enormous strength and confidence for people in any relationship. Maintain your faith in your team and your culture while everyone is remote. It will be too easy in a sudden shift to remote work to have moments of anxiety where you become insecure, not knowing if people are working as they should.

If or when that anxiety comes up, stop and do a reset. Go back to your culture and your processes. With your check-ins, collaborative project boards, and updates, you will know where everyone is in their work. With your company culture, you know the shared values that everyone uses as the basis of their work. The employer's *fear* of productivity loss in the work from home environment is the real issue. Don't fall into that trap and endanger the mutual trust you and your team share.

While it is a source of strength, trust is a fragile thing. Once broken, it can be something that is very hard to repair. Any monitoring software that the company may use needs to come with clarity from leadership. Be absolutely transparent in how, why, and when it's being used.

Keep the Client and the Team Happy

Remember that if you want to empower your people to do the best they can for the company, then you have to do the best that you can for them.

That begins with the foundation of shared values and is then expressed through a culture of trust. These values should be noticeable in your engagement with teams and in one-on-one engagements. Engaging in this way will have a profound positive effect on the company which will be expressed in attitudes of the employees and the work that is produced for customers.

Remember the hierarchy of needs that humans have and how they build upon each other. Safety and security, love and belonging, and self-esteem all are impacted by trust. Higher levels of transparency and trust result in a stronger sense of security for the remote worker. They provide fuel that feeds the mental well-being of the employee while they are separated from the rest of their team.

Enhance these positive effects on individual team members by reducing the number of virtual meetings, and make the meetings that do take place more impactful. Guide the rest of the team in best practices for meeting with clients and teammates. Prior to meetings, eliminate both online and real-world distractions. Prepare yourselves well by checking connections and closing extra browser tabs and apps to eliminate lag. Prepare your space, your lighting, and yourselves. Your appearance, posture, and mindset should project your expertise and professionalism. Be aware of nonverbal cues and remember that body language makes up a large percentage of our effectiveness as communicators. Make use of virtual whiteboards and visual aids when necessary, to better carry your points across the digital divide.

When you are an attendee, you should be equally prepared so that you can be fully engaged. Attention and focus are important for meeting attendees. It is noticeable when some attendees are not engaged with the meeting. It's not enough to just show up. A lack of participation in meetings stymies remote team cohesiveness. Also be cognizant that ebbing participation in meetings may be a signal of feelings of isolation.

Be Mindful of Signs of Isolation

When your team is working remotely, take notice of their appearance, their demeanor, and their engagement levels so that you can be aware of signs of struggling. Changes in those things and in communication habits are possible signals that someone is having issues. Be mindful in your remote management styles and be proactive. It is important to catch isolation and possible burnout in early stages.

And please do not forget about yourself. You are not immune to burnout or isolation. Far too many excellent leaders carry the weight of company burdens on their shoulders and forget how heavy they can become. At times, some leaders deal with stress like a frog in boiling water: the frog is gradually boiled alive without comprehending what is happening to him. He is placed in a pot filled with room temperature water. As the stove is lit, the temperature slowly rises, but the frog is unable to detect the change and so he stays in the pot. However, if the same frog was dropped into a pot of water that was already boiling, he would immediately jump out to save himself from the danger.

Take note if your sleep patterns change for extended periods. Be aware if you start to have lowered motivation or stop having interest in activities that you would normally enjoy. If you begin to have issues maintaining focus or become easily irritated, you may be struggling. This is not something to power through. You have the ability to make changes and to adjust your own workload and work hours. In fact, you have the responsibility to do so when you start to see signs that you are struggling.

Remember that when leadership is affected negatively by the strain of stress, it unintentionally spreads outward to the rest of the team. Remind yourself that your own well-being matters. You matter. Please take the time to take care of yourself. Exercise is a healthy and effective way to deal with stress. Encourage physical fitness in those you lead. One of the best ways to do this is through leadership by example. Maintaining social bonds is also a strong way to help stave off feelings of isolation.

Stay Social

All work and no play is counterproductive; in fact, eventually it will become debilitating. Encourage your team to continue to engage in activities and discussions that are enjoyable. Be proactive and create communication channels and opportunities for social interactions. Or better yet, have the team create these channels to better facilitate impromptu chances for socializing.

Continue to celebrate the milestones and events that you would have normally celebrated in person. Birthdays, work anniversaries, landing a big client, or completing a huge project should be recognized even more so in the remote working world. Recognition of efforts and good work as well as consistent feedback can do wonders for group morale and individual self-esteem.

Unless there is a health issue, try to schedule in-person group events. That will help team members who may find it more difficult adjusting to being suddenly separated from their team. If that is not possible, then a group activity such as virtual lunches with food delivered to the home of each team member from the same restaurant is a great option.

We want to ensure deeper feelings of connectedness when the team is distributed. Just because people are out of sight, they are not out of mind, but they may imagine they are out of mind due to being alone at home. Insecurities can set in when folks are not engaged and those random watercooler moments disappear. Help increase the feeling of belonging and decrease insecurities by offering more transparency, more visibility, and stronger engagement. Make documents and digital assets easily accessible. Make people easily accessible through your technology. Have company leaders add in open office hours, and set your online presence as online and available when you are not doing deeper, more focused work.

Generations and Personalities

Different generations have various world views that were shaped by the major events of their lives and social-political changes that took place in their younger years. They also have different priorities due to their various stages of life. At the same time, extroverts and introverts have personalities that also react differently to sudden shifts to remote work. Be cognizant of these unique traits and points of view. It will help guide your leadership during transition periods and help you better manage your people.

Look for the common ground that they all share and use that as a rallying point to avoid a perceived illusion of any generational gaps, which will help prevent stereotyping. Some common needs are flexibility and the positive reinforcement of being recognized and appreciated. Keep an eye on people who may be more prone to burnout and becoming overworked. Be flexible enough to share the workloads by reassigning tasks when needed as well as suggesting time off and away from their computer screens.

Your people are your best asset, and a heathy company culture should reflect that importance in the tone in which leadership communicates. Listening is also part of those communications. Don't simply wait for feedback to come to you; actively solicit feedback, and then act on it.

This helps people feel that they are a valued asset and that they have a stake in the company.

Leadership Grooms Your Culture

A huge part of people feeling as if they are stakeholders in the company is the behavior of the managers. If your current company culture is not up to snuff, then change it. Understand that you absolutely can change your culture. It won't happen overnight, but it will happen. You must begin from the top down. Change the behavior of the leaders and the behavior of the managers and you will change the company culture.

Strive for authenticity in all that you do. That should be the driving force in your engagement—not just engagement with employees, but with anyone the company interacts with during the business day. That creates consistency and makes your authenticity actually authentic. You can't treat group A one way and treat group B another way. True authenticity is not something that can be turned on and off. Simply put, that's phoniness. Your company's values should be able to be seen in any communication, in every action, and in all engagements.

Being authentic creates the opportunity to have a culture of trust. You cannot have one without the other. By adhering to this mindset and having it flow through your actions, you will empower your remote workers to continue their work with excellence. They will become less vulnerable to the rigors of the isolation. They will be less likely to feel disassociated.

The Right People

The interactions with your people should make them excited to be a part of the company and should cause them to look forward to work the next day. But if someone does not share the values of your company or if they do not believe in the vison and mission that dictates the company's reason for being, then they are not the right person for your company.

Shared values lead to a shared desire for the success of the group. Without those shared values, the desire for success is limited to personal success. Employees will be playing a solo game and not striving for the success of the team. Bringing on people who are the right cultural fit for your company will help you and every team member maintain a high-level confidence in their company while remote. They will know what to expect from leadership as well as each other, and they will know what to do in unprecedented situations.

This also pertains to subcontractors, vendors, independent contractors, and temporary hires. They should all share your company values. Their work at some point, at some level, will be intersecting with your brand. When that happens, their work will either be consistent with your brand image or it will dilute it. Either way, it will impact both your customers and your employees in the form of feedback and levels of satisfaction.

Hiring the right cultural fit will give you greater control over the company quit rate. As retention goes up, relationships within the teams strengthen. The longer people work together in a healthy environment, the more invested they become in each other's success, in mutual goals, and in the goals of the individuals.

If you hire remotely or plan to hire people who will be fully remote, adjust your hiring strategies accordingly. Look for skill sets and traits that are attuned to the remote work environment and virtual teams. Being self-starters and having an aptitude to work independently but with a team thought process are important characteristics. Written and verbal communication skills will help them be able to collaborate well, as will their ability to manage their time. Their ability to maintain a good life and work balance will be important to new people being able to fully function and stay healthy when remote.

Be prepared to have all digital engagement with the new remote team members. Remember that you may never meet in person. With this being the case, all of your engagement must be reflective of the company values and your culture. Augment that digital engagement with SaaS human resource platforms and employee recognition tools to keep up with your HR needs and employee satisfaction.

The World Is at Your Fingertips

With today's technology, the understanding of the needs of virtual teams, and the ability to work from anywhere, it truly does not matter where you choose to work. What is important is *how* you choose to work. Thriving while working remotely is a highly achievable goal, but you cannot depend on the technology alone to carry you there.

Technology is a tool. How you use it is key to your level of success. Leverage that technology to the fullest with good habits and a healthy, engaging company culture. Each word typed into your communications to the team must express the values of the company. Authentic engagement must empower your employees to do meaningful and purpose-driven work that they can believe in.

Remote work technology combined with a culture that values people, trust, and transparency is a winning combination for every company. Use this combination with authenticity and you will set your business up to thrive wherever you may find yourself. When it comes to your goals, think big and be bold. Fortune favors the bold for good reason. Bold thoughts backed by bold actions can make really big things happen.

Index